Foreword

I first met Roger in a field on the outskirts of Manchester in the summer of 2017. At the time, I was a student attending my first activist gathering with People and Planet, a network of students campaigning for social and environmental justice. As we ate dinner in the evening light, Roger approached the group with a nervous but determined energy. He invited us to a nearby tent for a talk titled *"How I Got King's College London to Divest from Fossil Fuels in Six Weeks."*

I was intrigued. Most stories we'd heard at this gathering were of slow, gruelling campaigns that rarely delivered big results. But Roger promised something different—a direct, effective approach. It had been two years since the Paris Climate talks declared that fossil fuels had to stay in the ground, and yet little had changed. A climate denier had just been elected to the most powerful position in the world, and I was bitterly disillusioned. That night, I decided to hear him out.

What followed was the beginning of Roger's journey to deliver over 200 public talks across the UK. He laid out the dire reality of climate science and argued that only high-impact, high-sacrifice civil disobedience could turn the tide. That talk inspired me—and many others in the field that night—to act. Within a year, we would co-found Extinction Rebellion. We decided to title this foundational talk "Heading for Extinction and What to Do About It."

Roger's journey started years earlier with a mix of study and hands-on activism. While working on his PhD, he didn't just study social change—he threw himself into campaigns to test what actually worked. He pulled off one of the UK's rare successful rent strikes and stood with migrant workers against terrible contracts. He did it all with a mix of guts and creativity, spray-painting campus buildings, organising like mad, and even going on a hunger strike against King's College.

"The university, like every neoliberal institution, relies on its reputation to attract investment. Beneath their greenwashing slogans is just a shallow concern for their image. It's about the money, not the people. Our campaign exposed that bollocks and created a reputational crisis so intense they couldn't ignore us," Roger told me that night.

Hearing about that win gave me—and everyone else—a jolt of energy. After the talk, I stayed back and asked Roger how I could get involved. He mentioned a campaign in London against air pollution. At the time, I was studying in Bristol, but I promised to stay in touch.

One month later, I spray-painted my local Barclays bank to protest its investments in fossil fuels. I was arrested, which caught Roger's attention. Over the next year, we kept in touch, and I found myself diving deeper and deeper into his way of doing things. I joined the Rising Up community in Bristol and, eventually, strategy gatherings in Gail Bradbrook's living room. There, we debated the future of our fragmented activist community and pitched ideas for the next steps. Amongst several proposals, Roger presented a paper for pivoting towards a climate rebellion. It captivated the room.

I read his proposal repeatedly on the bus home and late into the night. A dilemma gnawed at me: I wanted to be part of this rebellion, but I was supposed to go on a year abroad for my studies. The tickets were booked, and my family had invested so much in my education. In the end, I had so much faith in Roger's plan—and such determination to make a difference—that I quit university to join this eclectic and experimental group. Together, we set out to lead a rebellion against the British state's failure to act on the climate emergency. Roger and I were arrested countless times, and he is now serving a five-year prison sentence for giving one of those talks that so inspired me.

For Ramon, for glory.

Contents

Foreword 9

Revolution...**13**
Citizens' Assemblies: The One Revolutionary Policy 15
Citizens' Assemblies: Revolutionary Confrontation with the Carbon State ... 18
Dialogue or Fascism: The Plan to Save All We Love 22
Changing the Guard in the Gas Chamber: The UK Election 25

Science **31**
The Mysterious Ways of Love - Viral Predictions of Mass Death in
    Phoenix, Arizona 33
Rising Sea Levels Will Kill 40 Million In Our Lifetime..................... 37
Counting The Dead: Elite Pathology and Illiteracy 40
Billions of Lives on The Line - Existential Crisis Realities 46
The Truth About Atlantic Ocean Circulation Collapse 51
When Will Experts Start Saying "Mass Murder"? - Existential Crisis
    Realities November 53
Existential Crisis Realities: The Climate Science of October 2023 58
XR Turns 5 - Reflections on Telling the Truth 60
It is Better to Die With Honour Than To Die Like Sheep 62

The Left..**65**
The Information Illusion... 67
Here's Why the Greens Flopped in the EU Election......................... 69
The Intellectual Left Are Fucking Bollocks On "What to Do" 71
It's Time For Prophetic Leadership – Interview.............................. 73

Liberals **79**
A Primer on Reform and Revolution: An Open Letter to Green
    Party Members 81
The Guardian Facilitates Genocide 87
Labour's U-Turn on Citizens' Assemblies Shows They Fear the People.......... 92
A Small Matter of Treason: Starmer and The "Climate"................. 94
On The Farmers Protests 96

The British Establishment is A Cult: My BBC Interview 98
Visioning Extinction Rebellion 100
The Liberal Class is Complicit in Mass Murder 104

Conservatives **107**
Hope Against Hope: Recruiting Conservatives 109
System Change: Chatting with Conservative MPs 111
Civil Resistance: Between Protest and Just War 116

News Commentary **119**
Labour's Election Policy? To Shatter The Young's Bones For Eternity 121
Political Violence and Responsibility 124
COP 28 – Run by Fascists. Legitimised by Liberals 126
Just before the Beginning of the War 129

Spiritual **133**
Telling The Truth: The Easter Spirit 135
Finding Our Religion 137
Spirit = Action 147

Civil Resistance **151**
Why Repression is the Best Thing Since Sliced Bread
(Analytically Speaking) 153
Life's Trials are all Heaven's Pearls 157

Prison **159**
They Came For Us and Soon They Will Come For You 161
My Trial Was A Sham 164
Is There A Time Limit Or Not? A Letter To Celebrities 168
Who's the Real Fanatic? 173
Integrity vs Expediency: The Climate Trial 177
Law - What is it good for? 181
Jailed for Telling The Truth 182
I'm out of prison and mad as hell 184
After Despair: Awakening to the Revolution in 2023 187
English Gulag 191
I'm out of Prison! 196

Final Comments ....

    Beyond the Big Lie: The Return of America's Class Struggle ....

    How to Stop Fascism ....

    Hothouse Holocaust: Awaiting The Glory of Resistance ....

    Why We're Going Extinct - 5000 Fucks, 30 Million Views, 8 Billion De

    Reasons To Be Cheerful ....

Afterword ....

Acknowledgements ....

Final Comments **199**

    Beyond the Big Lie: The Return of America's Class Struggle 201

    How to Stop Fascism 204

    Hothouse Holocaust: Awaiting The Glory of Resistance............................ 207

    Why We're Going Extinct - 5000 Fucks, 30 Million Views, 8 Billion Dead ... 210

    Reasons To Be Cheerful.. 214

Afterword 217

Acknowledgements.. 223

This collection of 50 articles the corporate media wouldn't publish brings together Roger's most potent writings as he moved from rebellion to revolution. Rather than following a chronological narrative, it is organised around key themes: his battles with reformism, critiques of defeatism within the left, and reflections on spirituality and the transformative power of citizens' assemblies. His tireless energy and refusal to back down have fired up hundreds of thousands—and sparked plenty of controversy along the way. Roger's focus on microdesign, assemblies, and disruption has left a mark on activism everywhere. His ideas will push you to rethink what's possible in a collapsing world—and inspire you to imagine how we might rebuild it together.

- Robin Boardman, 27th of February 2025

Revolution

It's inevitable. Here's why.

Valerie Brown

Citizens' Assemblies:
The One Revolutionary Policy

(March 14, 2021)

Valerie Brown is standing as a candidate for London Mayor this May. Her campaign revolves around a single, transformative policy: the establishment of a legally binding London Citizens' Assembly to decide how to implement an emergency decarbonisation of the city. Alongside this, similar assemblies would address the critical issues and challenges faced by everyday Londoners.

This policy is revolutionary because it doesn't seek minor, meaningless concessions from the political system—it seeks to change the system itself. The system, as so many now recognise, is corrupt. To restore trust in democracy, it must be overhauled. For decades, conventional politics has proven catastrophically incapable of tackling the two defining crises of our time: inequality and the climate emergency. Politicians are trapped between the demands of voters for real change and the veto power of global corporate interests, which have no intention of allowing structural reform. Inevitably, they side with the rich and powerful.

The consequences have been devastating: mass disillusionment with politics, a widespread belief that change is impossible, and the dangerous allure of far-right parties peddling delusions of security. Meanwhile, progressive and left-wing parties cling to a political system that has repeatedly failed to deliver either social justice or effective climate action.

If the human and compassionate values built over generations are not to be swept away in the coming decade, we must take a deep breath and face the reality: it's either constitutional revolution or certain defeat. The solution lies in upgrading our political system to empower assemblies of ordinary citizens to make critical decisions about the direction of our society.

Why Citizens' Assemblies Work

There are two key reasons why Citizens' Assemblies are uniquely effective in addressing the crises we face:

1. **True Representation**
   Unlike any voting system, assembly members are selected randomly from the population. For the first time in history, decision-makers would genuinely reflect society. Half of the members would be women. Minority groups would be guaranteed inclusion. Most would be ordinary working people from across the country, while the wealthiest 1% would hold only 1% of the seats. This single innovation eliminates the systemic exclusion of the poor and marginalised from power. And because members are chosen by chance, no amount of money or influence can determine who participates. The rich would lose control.

2. **Deliberation, Not Division**

   Citizens' Assemblies base their decisions on deliberation. Members hear expert witnesses, engage in extended small-group discussions, and are free from outside interference or veto. The "miracle" of this process is that people quickly come to see one another as human beings, not political opponents. Through listening, understanding, and hours of discussion, they arrive at solutions grounded in collective wisdom. Compare this to the toxic soundbites of social media or the corporate-controlled mass media. Assemblies produce decisions driven by the common good, and the evidence from the past 20 years shows they consistently yield sensible, practical outcomes.

Not only do Citizens' Assemblies deliver justice and common sense, but they also foster a profound sense of community and democratic pride. Instead of a career politician delivering soundbites for the cameras, a builder from Birmingham or a care worker from Sheffield would speak on behalf of the assembly. People watching would think, *These people are like me*. This is what real democracy looks like: rule *by* the people, *for* the people.

In the face of the challenges ahead, we need a way of doing politics that unites us and helps us agree on what needs to be done. Citizens' Assemblies are the answer. They represent the Democratic Revolution of the 21st century.

This is why Valerie Brown deserves your support. Let's spread the word: *We can do this ourselves.*

Peter Bruegel the Elder, Fall of the Rebel Angels (1562)

Citizens' Assemblies: Revolutionary Confrontation with the Carbon State

(September 24, 2023)

"We have worked really hard over the past 14 years to build a reputation for facilitating robust and independent processes using a tried-and-tested model. We are known as an organisation that trusts citizens to draw meaningful and challenging conclusions when presented with a wide range of perspectives. To achieve this, we must be seen as impartial and not predetermining the issues or solutions. I have spoken to my fellow directors, and we feel that an organisational association with (you) may challenge our ability to do this in the future."

– From an email I received from a prominent
democracy organisation.

The periodic suicidal corruption and disintegration of elites is as old as the hills. A defining feature is the zombification of regimes, lingering for years after they are culturally and morally dead. These regimes cling to power in their final stages through a ruthless monopolisation of economic and political influence.

One primary mechanism in the endgame of our carbon regime is the patronage of state-supported sectors. To sustain a high-carbon, extractive, and ultimately murderous Western lifestyle, organisations must secure funding from carbon-state agencies, submitting to their core ideology of supposed objectivity: "independent processes", being "impartial", and "not predetermining the issues". These terms are designed to prevent structural challenges to the hyperobject—the ongoing destruction of organised human life through continued carbon emissions over the next decade. The 1.5°C threshold is now locked in, yet the social catastrophes this will unleash remain unspoken.

A second mechanism is the enforcement of chronic political passivity, driven by the mafia-like threats of reputational destruction from billionaire-controlled media—a phenomenon we might call the "Ed Miliband syndrome". For organisations tied to the regime, even the faintest association with progressive revolutionary movements aiming to save civilisation through civil resistance risks financial security and social privilege. The notion that disobedience can succeed—recent examples include Gary Lineker versus the BBC and Dale Vince raising funds for Just Stop Oil despite the *Daily Mail*—only underscores the psychic rigidity of these organisations.

In the so-called "democracy space", a small industry has emerged during the neoliberal era—the last 14 years—steeped in this insidious ideology. In practice, this industry facilitates the public relations agenda of preventing material challenges to the carbon-state project of mass death. Central to this agenda is the creation of Citizens' Assemblies, which feign democracy in three key ways:

1. **Carefully Curated Agendas**: Assemblies explicitly exclude discussion of the raw social impacts of carbon emissions—no sessions on starvation or the prospect of cannibalism, for instance.

2. **Broken Promises**: Outcomes are pledged to be acted upon by politicians but are systematically ignored, displaying the same contempt for democratic accountability seen after Keir Starmer's Labour leadership campaign.

3. **Sabotaged Visibility**: National publicity for assemblies is undermined, ensuring they remain misunderstood or unknown to the public—a far cry from the vibrant, participatory deliberations found in historical examples.

Yet, the *ideal type* of a Citizens' Assembly is inherently revolutionary. The carbon regimes, in their arrogance, make the classic collapsing-elite error of believing they can control this subversive democratic form. Citizens' Assemblies are a Trojan horse. People sense the pathway to their liberation. History shows us that pre-revolutionary periods are marked by autocracies attempting to use democratic institutions as distractions to prolong their rule—Charles I's Parliaments before the English Civil War, Louis XVI's Estates-General before the French Revolution, and the Russian Czar's Dumas during World War I. Today, Western governments are employing Citizens' Assemblies in the same way.

The plan is that these assemblies will serve the regime, but they have a glorious tendency to backfire. "The mob learns how to reason," as the anthropologist, David Graeber put it.

This is the reality of 2023. A new revolutionary objectivity is emerging, one that opposes the neoliberal version of "objectivity". The zombie carbon state will soon collapse as civil society uses Citizens' Assemblies to create alternative governmental institutions—undermining rather than propping up the regime. We won't ask the regime to organise these assemblies; we'll do it ourselves, without their "experts". This changes everything.

These assemblies will expose the naked truth: politicians can no longer deliver on their promises of decent living standards as extreme inequality and climate extremes drive fiscal crises. They will continue "maxing out" emissions, leading to the destruction of billions of lives. The carbon regime will be revealed as the greatest force of evil in human history. Meanwhile, the liberal administrative classes—the house slaves of this genocidal system—will try to dampen the revolt. "We feel that an organisational association with you would undermine our ability to be impartial," they'll say. But they are already finished. Times have moved on.

The assemblies of resistance will confront the monstrous denial of moral truth. Participants will challenge the pervasive irrationality of separating physics from society. They will refuse to accept the sinister abstractions of terms like "a liveable future" and confront the harsh realities: mass slaughter, rape, starvation, and the grim prospect of bodies piled against endless miles of barbed wire.

After their deliberations, they will declare the greatest moral message of our times: **Never Again**.

The New will finally be born, and the Dead will finally be laid to rest.

Art: Time Door Time D'Or, James Rosenquist, 1989

Dialogue or Fascism: The Plan to Save All We Love

(Feb 2, 2024)

You wouldn't know it but millions of people are going to experience a lot of physical pain - not to mention the emotional pain of seeing the physical pain of those they love. The source of this pain is something called the Atlantic Meridional Overturning Circulation (AMOC). This is the Atlantic current that comes up from the equator and keeps Europe warm. Last year it was revealed that Europe will experience an overnight fall in temperatures of 3-8C at any point after 2025 due to the AMOC collapsing. Why? Because the Greenland Ice Sheet and other Arctic ice sheets are irreversibly melting.

This prediction is an underestimate. Predictions by climate scientists are conservative because they only look at factors which are certain - and ignore uncertainties, even though they are very likely to happen. And so, for instance, it is no surprise that a few months later we have been told that the Greenland melt has been underestimated. We are now told this every year. As long as a defective methodology is used by scientists we can look forward to their predictions getting exponentially more conservative over the coming years. "Scientists were shocked" and similar phrases now occur in most "climate change" articles.

And so we are heading to fascism. The first reason for this is because of the refusal of the liberal and mainstream media to make clear that "climate change" means physical pain - lots of it, and then you will die. The collapse of the AMOC is just one of several catastrophic situations which are now locked in. But the real catastrophe is the unwillingness to acknowledge these are social crises, not "climate" crises. What needs to be communicated is what it will mean for your pension (it will not get paid), what it means for your food supply (it will collapse), what it means for law and order (you are likely to get shot).

The fact that such predictions seem outlandish is due to the greatest scandal of our time: the refusal of the media to tell people what ACTUALLY is going to happen. For instance the AMOC collapse will produce an overnight reduction of average temperatures by 3-8C in Europe (closer to 8C in Northern regions). Without a shadow of a doubt this will destroy our prosperity. At a minimum we will be thrown into a World War Two emergency - a state command economy and compulsory rationing. Think of an eternal Covid Lockdown and then times how bad it is by 10. That gets you in the ballpark of what's ahead of us.

Why don't people know this? Because no one in the media brings in an agricultural expert to tell us what an average -5C temperature situation across Europe means. It's obvious. It means mass starvation unless the country comes together into a wartime mode.

And if we don't all come together? Then we will experience the biggest shitshow in the last 10,000 years of human history. Wave after wave of crises which will get compounded into holocaust episodes as fascistic forces drag us into a hell of permanent war. Think about Gaza. And then think about it again. That is what happens when fascism takes hold. And it is taking hold all around us - wake up!

Policing Just Stop Oil

Changing the Guard in the Gas Chamber: The UK Election

A new party is in power, but it's the same old story. Only a revolution will free us from society's suicidal prison.

(July 22, 2024)

Note: *This was the last article Roger wrote before being sentenced to 5 Years in Prison.*

If you go to prison (or should I say when?), you will get a clear understanding of what a guard does. They enforce rules. You can walk around your cell - fine. But walk around the prison yard after time is up and they drag you inside. You will get a few scrapes and bruises (or worse) and then be put in an isolation cell as punishment. There are many rules, but only one main rule - break the rules and you will be punished.

It's the same with politics at the end of the world. You can vote for whoever you want to as long as they don't try to stop the project to destroy the human race over the next two decades. To quote those hysterical idiots at the UN, we have "two years to save the world" or economies will be "devastated".

Of course, saying something revolutionary like this has been called an ideology for a good 200 years now. In fact, the idea that the end is coming has always been part of human culture. An ideology is the imposition of an idea upon reality. More often than not the ideology is actually contradicted by reality - the end of the world is not coming. But not always. Read that again - *not always.*

The exception does exist. It is there in the historical record. Civilisations do destroy themselves - they "commit suicide" to use historian Arnold Toynbee's famous phrase. Actually, they all destroy themselves eventually. Nothing lasts forever it seems. The cruel paradox is that they destroy themselves in large part because they are so sure they will not destroy themselves. Sometimes the "ideology" is the opposite view: the ideology of "progress" - the imposition of the idea that the end will never come.

There are, however, facts. Things which exist independent of our beliefs - as I am trying to inform the judge in my trial this week. There is no chance he will accept it as his ideology trumps reality. The facts are that capital controls the world in the 2020s. International capital, to be more precise, has escaped control by the state. This is not an observation - it is an empirical statement. It is historically situated - now in this decade. As such it is an ideological statement. Capital does not always trump the state - in fact usually in history, it does not. And soon capital will lose again.

Roger's sentence was on the front page of 4 major British newspapers. All focused on the details of the trial rather than the existential threat the defendants were trying to avoid.

Roger's sentence was on the front page of 4 major British newspapers. All focused on the details of the trial rather than the existential threat the defendants were trying to avoid.

The deal is this. If you try to interfere structurally with the central character of capital - the ability to engage in economic activity that externalises the social and ecological costs - then capital will punish you. Tinkering is fine - you can always linger a while before you're forced to leave the prison yard - but a structural challenge to capital's ability to cut costs will be punished. You get capital flight and increased debts to a global power - just look at Greece after the 2008 financial crash, they lost €4 billion a week as they resisted a European bailout. Your country will be poorer and you will be voted out of office and then someone else will have a go at squaring the triangle.

What is the endgame here? There are two outcomes. The state collapses, hollowed out by the social costs of capital's externalities, which leads to social collapse. Or the state forces capital to submit to its will through its monopoly of violence. Do what we say or we will put you in prison (or worse). In this case, the boot is on the other foot. Both endgames result in what everyone was trying to avoid - getting a lot poorer. "Make a stitch in time to save nine", but you know what dear old humans are like - they always end up making nine stitches. In universities, they call this the collective action problem.

What's the upshot? Whoever you vote for, you will get the same outcome - the rule of capital. Say this to half the population - maybe it's only a quarter now, and soon it will be a tenth - and they will tell you you are mad. Tell the growing other half and they will say "Yeah, obviously". It's not like people don't know the score. It's just that these people never get to speak in the public sphere - in the media. That's another one of the rules.

Then there is the gas chamber. "Oh that Roger is always talking about the Holocaust". To which I reply "No comment" (I have learnt that rule!). It's you making the comparison, not me. So maybe it's you who is being "anti-semitic" for merely thinking that there could be a comparison. This is how ideologies finally eat up their children, by the way. The very awareness of system critique becomes a crime itself. The accusers become the victims. The guards are punished for guarding. Think about the great show trials.

No, what I am saying is that the world we literally live in is a gas chamber - I am making no "comparison". That is what it is. We are surrounded by gases and these gases are confined in space. Go outside this space - by digging too deep or climbing too high - and you run out of oxygen and die. Life depends on that small sliver of habitable space, the Goldilocks zone.

Let me illustrate this by telling you about my farm. My farmhouse is situated 150 metres above sea level. I can grow food from 100m to 200m. Above 200m it is too cold and windy. Below 100m is too wet and swampy. We live in a sliver - around 100 metres thick - and outside of that we starve. It's this sliver that also exists within the gas chamber.

Capital has been externalising greenhouse gases to the point that it is significantly changing the composition of this gas chamber. These new gases will not allow you to grow food reliably at scale, even within the sliver. As we will soon find out, no food means no life. In other words, unfettered capital has created the endgame of all endgames - extinction.

Science

1.5C is dead. Here's what we face.

Newton, by William Blake. Blake believed that Newton's scientific approach to the world was too reductive. Here he implies Newton is so fixated on his calculations that he is blind to the world around him.

The Mysterious Ways of Love - Viral Predictions of Mass Death in Phoenix, Arizona

Jun 6, 2024

This week one of my tweets, predicting the mass death of 12% of Phoenix in the 2030s, went viral, accumulating 1.6 million views. Several scientists have pushed back on it with one chief scientist at Berkley even bluntly dismissing it as "dumb". Here's why their twisted love of certainty is actually a form of denial.

In the film Beautiful Mind the main super nerdy guy, John Nash, goes up to a woman in a bar and tells her the truth - he wants to have sexual intercourse with her. In a predictably Hollywood way, she slaps him across the face. Sometimes the truth is not actually the whole truth - there is another sense of truth which is more appropriate, more real. When you go to a funeral it is "true" that the grieving partner will "get over it" in time. But you don't go up to this person in the depths of grief and say, "Ah there's no point being down - cheer up. Plenty more fish in the sea". Only a dick does that.

The problem with scientists (with the unique exception of NASA's Peter Kalmus) is that they think like John Nash. They approach women like they approach maths. They communicate about "climate change" as they communicate about moss. I had a scientist friend years ago - super nerdy - who studied moss. He was an international moss expert and explained the ins and outs of moss in a highly competent, professional way. But no one gets emotional about moss. The "climate" is entirely different - it is jam-packed and overflowing with emotional and existential angst. In a reductionist way, of course, climate is a physical thing just as moss is a physical thing. But "things" don't exist in our consciousness - only "things-in-an-emotional-context". Understanding this is vital for our culture if we are going to survive.

Let's look at another analogy: think about a fire starting in your house. You run to the neighbour and scream "My kids are going to fucking die". If a scientist was to say "Well that's not quite right - you need to do a proper risk analysis - papers suggest that the death of children in fires..." that would be really fucked up. And so is you running to your neighbour and saying, "My house is on fire with a 55% chance of burning down and a 30% unsubstantiated probability of severe burns to my children (who are almost certainly mine) - would you be so kind as to come and help". It would be totally fucked and inhuman.

But this is how the public sphere expresses itself on "climate" - that technical repressive word for the prospect of the agonising hell of slow death for ourselves and our children.

Heat-releated deaths in Maricopa county, Arizona

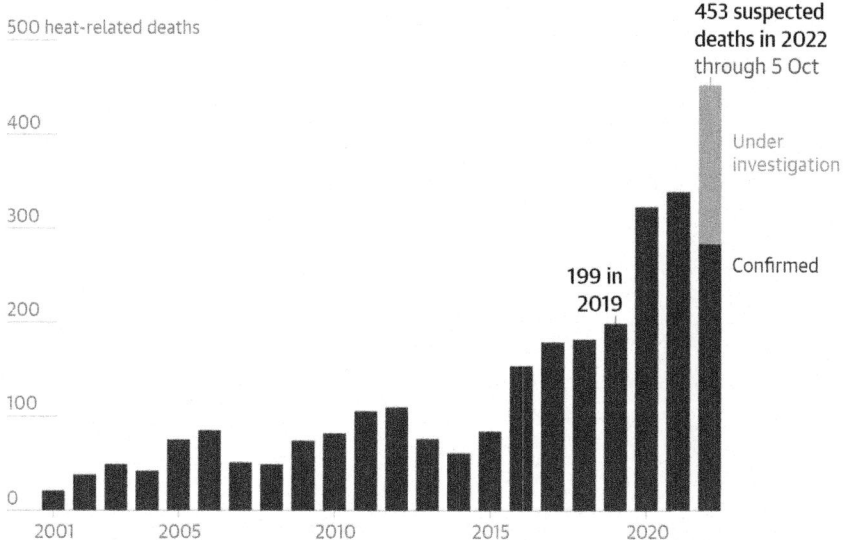

Guardian graphic. Sources: Maricopa county, office of vital registration and office of medical examiner, Arizona department of health services.

Last year, Phoenix suffered a month of consecutive days over 110F (43C) and a record 645 heat deaths – a 700% rise over the past decade.

The nerdy scientists are the mirror image of the raging deniers. They feed each other. They both need each other like one of those really perverse dependency relationships. If scientists emote and speak in blunt terms like "my children are going to die", the deniers give them hell for saying things that are not certain. "Have you proof that your children will die you fucking bunch of fear mongers? No!". And so the scientists retreat, which suits them just fine because then they can continue to be in their comfort zone of only talking about certainties: "Sorry you are right, there is only an uncertain possibility my children will soon enter into unique levels of discomfort".

Both scientists and deniers are not computers - they are humans trying to avoid emotional pain. Both groups can't cope with me because I am the guy running into the room going: "I don't give a flying fuck about your fucking certainty analysis - it's my fucking children you total dicks." That's why I got 1.6 million views. Because I am violating the rules both sides impose on the public sphere. That is why my favourite historical figure, Larry Kramer, was such an effective change-maker. Because he didn't give a fuck.

This is a roundabout way of saying that when I say, "12% of Phoenix are going to die" I am not really saying 12% of Phoenix are going to die. I am not talking about moss - I am making an emotional, rhetorical and entirely appropriate statement. A deeply true statement in the holistic sense.

Art: Paula Rego, Flood, 1996

Rising Sea Levels Will Kill 40 Million In Our Lifetime

(Mar 22, 2024)

It has recently come out that the rate of ice melt on Greenland has increased by eight times (yes that is right, *eight* times). I have yet to speak to the relevant scientists involved, but it seems reasonable to assume that it has more than doubled every decade.

Why does no one do the social science on these numbers?

I'm a social scientist. More specifically, I did Phd research at King's College London for five years. More specifically still, you might say I learnt how to do maths about society without fear or favour. Which, as you can see below, is a radical thing to do.

So let's get going and do a bit of maths...

This decade we are looking at 12,500 cubic kilometres of melt. Given the latest report, we should double it each decade from now on. Greenland has 2.9 million cubic kilometres of ice. So, effectively, the whole of Greenland will have melted by around 2080. That's within the lifetime of people in their 20s or younger today. That will raise sea levels by seven metres.

To get a flavour of the geophysical amplifiers involved watch this video on Youtube by two expert professors: **_Greenland: Ice Loss Accelerating_**.

Taking into account the melt of other areas of the world, such as Antarctica, we can estimate that effectively everyone living within 10 metres of present sea levels will have to move. That would be 600 million people today - more like 750 million, given an increase in the world's population to 10 billion from today's 8 billion by the mid century.

In a conservative estimate this will lead to 40 million deaths in war (5% of the total refugees), given the "war" will continue forever (i.e. the sea level will continue to rise). It will lead to 20 million women being raped (look at the relationship between those killed in social breakdown/war and rape in places like Congo, Rwanda, Nanking - it's around 50% of deaths).

Add in the numbers already established for temperature rise (extreme heat) - 1 billion refugees at $2°C$, 2 billion at $3°C$ of warming.

These numbers do not include the refugees caused by sea level rise.

So let's say **2.5 billion displaced people towards the end of the century - one in four people on the planet - 125 million deaths in war/social breakdown and 60 million raped women.**

And the most important point: it does not stop - the ice continues to melt. This is nothing compared with what the 2100s will bring.

This is the main scenario for this century with our present knowledge. It's 50% likely to happen with a 25% chance of doubling because of the compounding effects of endless social collapse and fascistic dysfunctionality (things did not end well for Germany in 1945). And a 25% chance of these numbers being halved if there are democratic revolutions which create rational policies to slash emissions and do geo-engineering (though sea level rise is now locked in regardless of what happens to emissions - it's warm: ice melts).

What to do about it is a question only those who are still in denial ask. It is beyond obvious to the 1% of the population with "self knowledge", as Timothy Synder calls it. Give up your job and dedicate your life to enacting revolutions. So that when all this shit happens you haven't got the additional shit of knowing you stood by and did nothing effective to stop it.

The Great Day of His Wrath, John Martin, 1851

Counting The Dead:
Elite Pathology and Illiteracy

(Mar 9, 2024)

Recently David Wallace Wells, author of **The Uninhabitable Earth,** wrote a nice little article in the **New York Times** which challenges my use of "one billion climate deaths". The figure comes from the peer-reviewed <u>Pearce paper</u>, which states:

> "If global warming reaches or exceeds two degrees Celsius by 2100, mainly richer humans will be responsible for the death of roughly one billion mainly poorer humans over the next century."

The New York Times article, "Just How Many People Will Die From Climate Change?" tells you all you need to know about the methodological illiteracy and political pathology of the global north elites. Put simply, they can't think straight and they don't give a shit about killing poor people.

Elites Can't Think Straight

1. This is what natural scientists do on the elites' death project. They go: "Factor X has only a 25% chance of happening - so we can't be certain - so we can't put it in the model." Meanwhile, the rest of the world does a standard risk analysis: "How many people will die if X happens?" and times it by 0.25. So if X will lead to an estimated 100 million deaths, you put in 25 million. You don't ignore that.

2. They also don't tell you the time periods. Notice there are no time periods in his article which is analytically ... well, dumb. But it creates the subliminal, deluded message that the death project is an event that happens and then stops. If we're talking about how many will die from climate change, what's your time period on something that will get locked in for 100,000 years (at least)?

Wells says: "I don't think it's right to suggest that reaching two degrees of warming (which now looks very likely) will mean a billion people dead."

Ok, so let's do some actual numbers here. First let's define what a $2°C$ warmer world is. Five years in a row that average $2°C$ or above warmer than pre-industrial levels seems reasonable. This would then lead to a triggering of all the major tipping points: Amazon droughts, permafrost melts etc. That would then take us over $3°C$ of warming within a generation, regardless of what human emissions do.

Peer reviewed predictions are that two degrees average temperature increase will result in one billion refugees. If fascist states breakout in response to such social breakdown, as they often do, then it's reasonable to expect that 10% of that number will be killed in the process and 5% raped (100 million and 50 million). Check out recent history in The Congo, Rwanda, Cambodia, Nanking and the Soviets in Berlin for starters. Other papers say that three degrees will lead to two billion refugees, which is currently the most likely temperature increase for the end of this century. If we follow just the logic of social collapse, that's 200 million deaths.

You can see how a proposed **New York Times** article in touch with reality would get rejected even at this stage, but let's stride on ...

Let's add in secondary effects. In his piece Wells questions my use of "a somewhat obscure paper". So for this I'm going to rely on Professor Nicholas Stern, Chair of the Grantham Research Institute on Climate Change and former Chief Economist of the World Bank.

In 2006 he and his team wrote the 662 page **Stern Review** for the British Government. Probably the most comprehensive economic study on the impact of climate change to date.

In it he states: "Our actions now and over the coming decades could create risks of major disruption to economic and social activity, on a scale similar to those associated with the great wars and the economic depression of the first half of the 20th century...Hundreds of millions of people could suffer hunger, water shortages and coastal flooding as the world warms."

So, get your pen and paper out. Estimates of military and civilian deaths in World War Two range from 70–85 million. That's about 3% of the world's two billion population in 1940. At the current eight billion population that would scale up to 240 million dead. Estimates of deaths caused by World War One hover around 20 million deaths out of a 1.8 billion global population. So a similar scale today would be around 100 million deaths. Estimated deaths from the Great Depression are up to 10 million, when the population was a quarter of what it is now, so that's around 40 million today.

You get 300+240+100+40 = 680 million ... just from heat, war and economic depression.

Okay, now add in water shortages, disappearing glaciers, methane bombs, water vapour, arctic blue ocean event, AMOC collapse, coral reefs, ocean heating (the list goes on)...

To give you a number, Dr Ripple recently published a report alongside Sir David King that stated:

> "By the end of the 21st century, **as many as 3 to 6 billion people** may find themselves outside the Earth's liveable regions, meaning they will be encountering severe heat, limited food availability and elevated mortality rates."

Is anyone serious that the main scenario does not take us over one billion deaths, with a thick long tail probability of burn baby burn fascism taking us to effective extinction (i.e 6-8 billion deaths)?

They Don't Give A Shit About Killing Poor People

Wells also says the paper I quoted for this figure is "focused less on climate impacts than on climate justice."

Because "climate impacts" have no connection to "justice"? - Bit of a revealing slip up there!

The article contains all the usual bystander words: "shocking", "scary", "worry", "speculations": the euphemisms the elites use when they kill the powerless throughout history.

And then there is the classic displacement title: "Just how many people will die from climate change?"

People don't die from "climate change". They are killed by those who make political decisions that mean disposing of people is worth it, to keep their power and privilege. Concentration camps do not kill people - they are killed by those who put people into the camps, right?

And finally, there's that deeply revealing first 'just' word in the title. "Well, what a fascinating question, Sir, – 'just' how many deaths of those poor coloured folks are we looking at here? Such an interesting subject..."

Only a culture completely devoid of empathy could write such a word in an article about such obscenity.

The elites' pathological inability and unwillingness to follow through on the logic of even their own statements is particularly evident in Well's case. He's written a book called *The Uninhabitable Earth*. It's good stuff. So, sorry to ask the question, but what does uninhabitable actually mean?

It means: "unliveable. unfit or unsuitable to live in or with."

So let's take this nice and slowly then.

"Unliveable" means you cannot live and therefore you die.

If you die then you are...dead.

The earth is where everybody lives.

There are eight billion people on the earth

So...

8 billion people are going to die

That is what "uninhabitable earth" actually means.

I feel like vomiting. So, I will stop there.

Art: Los Moscos, Mark Bradford

Billions of Lives on The Line - Existential Crisis Realities

(Feb 17, 2024)

On BBC's HARDtalk 5 years ago, I was slammed across the board for saying billions of lives were at risk. The liberal climate movements tried to refute and play down the existential risks ahead. The last few months have shown ever more clearly that I was right. After all, what do we actually think the "end of civilization" will look like?

Several new scientific papers from Ripple and Pearce have outlined that not only are billions of lives on the line, but that "richer humans will be responsible for [the] killing". This month, the NASA Climate Scientist, Peter Kalmus, also added his voice to that position. The time for shying away from our existential reality is over. The precautionary principle dictates that these realities form and guide the mission of our lives - that only a revolution to change who makes the big decisions is going to save us.

---

*"If left unchecked, every year on average will be hotter than the last, and at some point ... global heating will take down civilization as we know it. **Billions of lives are at risk ...**"*

- Peter Kalmus, NASA climate scientist.

Maybe 2023 will be remembered as the year when the word "billion" entered the public sphere as the number of people who are going to be killed. When the liberal climate industry breaks down and people start to focus on just how serious the situation is.

Exponential Temperature Increase

The situation is that temperatures have been going up exponentially since the 1970s and we will pass 2C in the next 15 to 20 years. After which feedback loops will be triggered and we will continue past 3C and 4C.

Here's just a snapshot of what those exponential temperatures really mean:

"By 2050–2074, two successive years of single or compound end-of-century extremes, unprecedented to date, exceed 1-in-10 likelihoods"

Such extreme temperatures mean that one in every ten years our ability to feed ourselves will be destroyed. Don't look away.

Fossil Fuels Are At A Record High

Despite years of commitments from countries to slash the emissions of greenhouse gases that are warming the planet, they are still on the rise. Carbon dioxide released from burning fossil fuels is expected to rise by

1.1 percent in 2023 compared with 2022, scientists found in an extensive peer-reviewed analysis published this week.

This is what it looks like when cognitive dissonance goes non-linear. Especially when they have even got the full picture.

> *"Emissions have been soaring well beyond official statistics reported by countries."*

On top of everything else, we find out they have all been lying as well. Are we really going to let these criminals take our children into an endless hell hole?

Worse Than Previously Thought

Climate Impact Research (PIK) shows that state-of-the-art climate models significantly underestimate how much extreme rainfall increases under global warming

"The intensity and frequency of extreme rainfall increases exponentially with global warming,"

So much for "state of the art" - Linear analysis is fucked. How many experts are in their own bubbles of expertise and not looking at the bigger picture? Cascading tipping point will send us into a hothouse earth. Forever.

Just look at this new report on ice levels:

If there is surely one phrase that sums up 2023 it is **"worse than previously thought"**.

Even the new Global Tipping Points report minces its words.

> *"The climate change assessment process has tended to focus on the most likely outcome, rather than evaluating the highest-risk outcomes. But this is poor risk assessment and it is leaving society ill equipped for what lies ahead."*

Understatement of the century – **it's a total fucking disaster!**

How many euphemisms do we have to deal with before someone finally breaks free of the system and the corporate media to say how it really is?

This is what the Labour leadership refuses to accept - there is a point when we cannot stop the chaos, and the chaos gets exponentially worse. Reformism is simply irrelevant according to the science. It's self-evidently a treasonous crime for Keir Starmer to pretend otherwise.

Economic Risk

> *"Nature is not the elephant in the room, it's the huge green scorpion running towards us"*

At some point the penny will drop that nature is stronger than us, especially at above 2C global average temperatures. Yet to add to the fuckery, that realistion is going to be seriously delayed by fossil fuels firms covering up the dirt, as this new report in Nature points out:

> **"16 of the world's most polluting fossil fuel companies were associated with more than 1700 climate misinformation ads on Facebook alone, with 150 million user interactions"**

It's clear that the public sphere is completely corrupted - the solution: door knocking, leading to assemblies, leading to revolutions, leading to legally binding citizens assemblies. Got a better idea?

Water Crisis in Gaza

The water situation in Gaza is catastrophic. Even before the war, Gaza had virtually no potable water. The population relied on a polluted and rapidly depleting aquifer, as well as a limited number of desalination plants that fell far short of meeting local needs. This has led to the alarming statistic that 97% of Gaza's water is unfit for human consumption.

One of the things no one wants to talk about is how many civilians will die once the whole of the Middle East faces social collapse as the region becomes "uninhabitable" (that nice environmentalist word for total fucking murderous hell), as the world's rich take us over 2C.

COP Fuckery

Part of me feels bad swearing away at the fuckery of COP28. 'Calm down Roger - don't get carried away.' Then we read the latest from Dr. James Hansen - arguably the greatest climate scientist of our time. What does he say?

We are going over 2C - meaning we face the prospect of 1 billion people uprooted and forced to migrate. No social scientist or historian with any self-respect can avoid the judgement that this will be total hell: 20 times the number of refugees produced by WW2.

Me swearing away on twitter should be the least and last of your worries.

I don't know if Dr. James Hansen will appreciate me saying this, but he is one of my favourite climate scientists. Not just because of his immense expertise, but because he is one of the only top scientists in the world who is capable of calling a spade a spade.

UK Temperatures to Plummet

More importantly, temps will stay below 10C in summer. As a farmer I can tell you there's not a chance in hell of being able to grow summer crops at scale in those temps.

What sort of total fucking stupid death cult political regime takes this risk?

Théodore Géricault. The Raft of the Medusa, oil on canvas. 1818-19

The Truth About Atlantic Ocean Circulation Collapse

(Feb 12, 2024)

As I write, an article on the Atlantic Ocean Circulation nearing its tipping point is the top viewed on the Guardian website. This is what they miss out:

1. The collapse of the Atlantic Ocean circulation (AMOC) will be the most devastating event in the last 10,000 years of human history.
2. It will happen overnight with sudden effects.
3. It will be irreversible and continue for 1000s of years.
4. It will destroy human civilisation because it will be impossible to grow food in northern Europe - temperatures would drop by

between 3-8°C. Enough to half the amount of land where you can grow wheat.

5. 100s of millions of Europeans will have to move or starve to death. Those that move will be subject to holocaust events created by warlords and/or fascistic regimes.
6. Coastal cities will have to be evacuated.
7. Monsoons in the tropics will collapse, resulting in 100s of millions more refugees.

This is just the beginning - the collapse also will feed into other disastrous climate tipping points like the collapse of the Amazon rainforest. We are looking at billions of deaths and possible effective extinction this century - that now has to be the main concern.

Last but not least, the above scenario is a conservative prediction because it doesn't take into account the non-linear effects of other systems on the AMOC collapse date (e.g the collapse of ice cover in the Arctic, methane release, and mega forest fires).

Why is no one talking about this?
Why aren't there emergency conferences of Europe's farmers?
Why aren't the media going on strike till the government acts?
Why aren't there mass sit-downs in cities for weeks on end?

Because repressed scientists just say that it's "kind of scary" - like saying Auschwitz was "kind of unpleasant".

The situation is totally fucked.

Scientists Gather Outside Parliament

When Will Experts Start Saying "Mass Murder"? - Existential Crisis Realities November

(Dec 14, 2023)

As the catalogue of impending disasters increases, journalists and experts are still refusing to call a spade a spade. November was another euphemistic month of 'Catastrophe' and 'Warnings' that were emotionally detached from the reality at hand - the mass murder of humanity.

How else can we describe the elite's continual endeavour to put carbon in the atmosphere, knowing it will wipe out life on this planet?

This monthly science briefing lays out the latest scientific evidence alongside the media's coverage of it, followed by my critiques. The aim is to update you on the science whilst also cutting through the liberal media's repressed distortions of the truth.

Daily Temperatures Breached +2.0°C For The First Time

Using different data, two of my colleagues @LeonSimons8 @EliotJacobson just broke the news that we have just breached + 2.0 C degrees. (CAVEAT: Breached once doesn't mean it's an average warming temperature.) But it appears life on Earth is under siege. https://t.co/pErHRnGohH pic.twitter.com/8eUTapwQTZ

— Dr. William J. Ripple (@WilliamJRipple)
November 20, 2023

So, we passed 2C. But it's okay, it was only for a few days. Back to work everyone. We're not going to get killed by the elites just yet.

This from Prof. Hansen - "The magnitude of the observed warming is off the scale" "within a decade or so it will probably be 2 degrees" Remember, 1 billion will face potentially lethal heat stress if global warming reaches 2C above pre-industrial levels. Thanks to @PaulHBeckwith pic.twitter.com/y9EjzOSPhs

— Peter Dynes (@PGDynes)
November 26, 2023

In just a few lines this sums up our situation. We're heading towards billions of deaths. Hansen's colleague, researcher Leon Simons, added "All this extreme ocean surface heat is about to turn into a year of shocking extreme atmospheric temperatures and even more extreme weather. We might be getting a first taste of the Termination Shock from terminating part of our cooling sulphur pollution, while still increasing greenhouse gases."

"The data suggests an alarming trend where temperatures continued to soar after previous Super El Niños, indicating an impending crisis. As you can see in the updated Climate Spiral, the climate seems to be spiralling out of control and sending us over 2C."

Note: We passed "Daily Temperatures" of +2C. The Climate Spiral shows "Monthly Temperatures" which are yet to exceed 2C. We generally talk about climate in terms of "Annual Temperatures" but these other figures give us the latest, up-to-date picture of where those annual temperatures are heading.

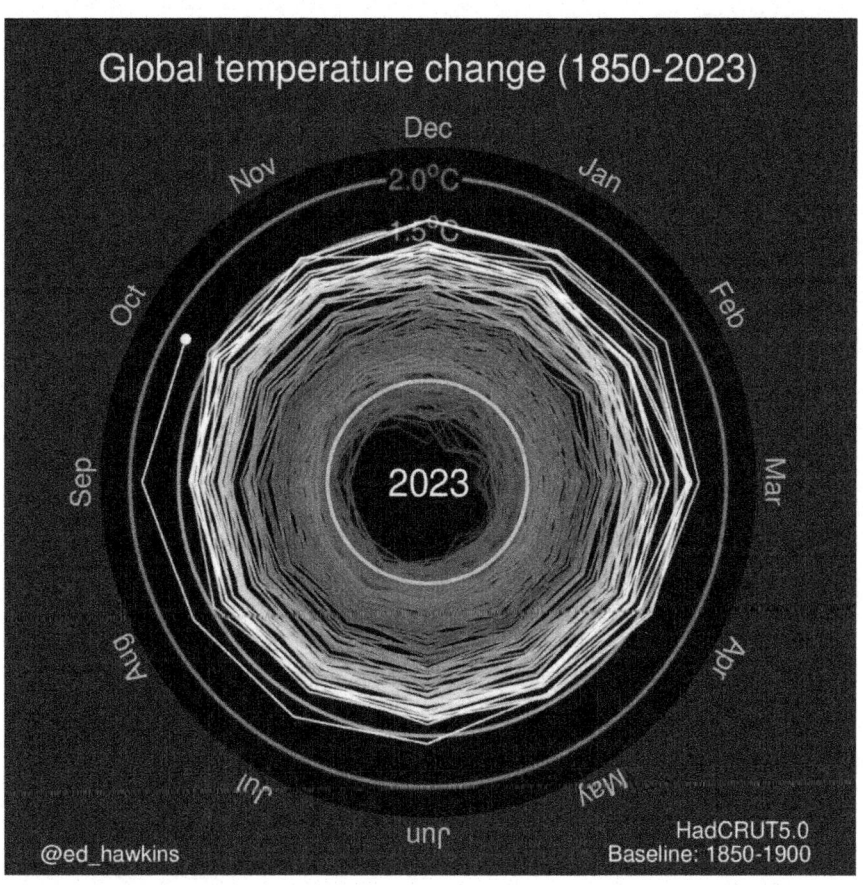

Carbon Brief's new report shows that a 2C world would also lead to polar oceans that are "ice-free" in summer and suffering "essentially permanent corrosive ocean acidification". The trend signals an imminent rise in global sea level "between 12 and 20 metres" of sea level rise "if 2C becomes the new constant".

This report shows the central contradiction in how the science community communicates "the facts". On the one hand it says if we get to 2C the situation will be "catastrophic" - implying ... oh, okay, so that's the deal then.

And then elsewhere it says that 2C will lock in "additional warming" from permafrost melts and other feedback loops. ... So there is no such thing as "2C". There is stopping at about 1.5C (if we are lucky) or speed up to 3, 4, 5C plus - whatever - meaning billions of people starve to death.

"Scorching heat highlights Brazil's inequality"... I would have thought it highlights the extreme sadism of global north elites and their indifferent publics that created the heat in the first place. Though highlighting the "glaring inequality" makes The Guardian readers feel virtuous in the "left" analysis, as a displacement for their glaring willingness to be bystanders.

During the heatwave they also reported on the "Death of a fan at a Taylor Swift concert on Saturday night in Rio de Janeiro, at which thousands of other concertgoers reportedly had to be treated for dehydration."

For the mathematically inclined, the question is how many more degrees should it need to reach before 1000s die and millions have to be treated for dehydration? It's a curve, remember.

So, unless a whole heap of scientific data from NASA satellites and ocean sensors is wrong, we are in for escalating troubles with Earth's climate in the very near future. We have to go on a war footing against this immense threat.

— Prof Nick Cowern (@NickCowern)
November 9, 2023

Prof Nick Cowern says "We have to go on a war footing"

"Human-caused climate change resulting in higher average temperatures has caused a global decline in snowfall, according to a new analysis from NOAA.

That means more precipitation is likely to fall as rain, causing **a disruption to food supplies and less water for billions of people**, the report said."

"Beyond less water and disruption to the global food supply, reduced snowfall means higher ground temperatures, too."

Note again the use of Billions. Billions of people will be "disrupted" from food and water supplies. What does that mean when billions of people can't access food and water? Oh yeah - they die.

They are going to be killed by our elite and carbon infrastructure will be the murder weapon.

Existential Crisis Realities:
The Climate Science of October 2023

(Nov 22, 2023)

October 2023 brought a whirlwind of revelations, warnings, and urgent calls for action from the climate science community. In this new monthly series, my team and I will look at scientists' updates to keep those in resistance in the picture and armed with the latest research.

6 billion people could be in unlivable regions by 2100

An international coalition of climate scientists, in a paper published on October 24, declared that Earth's vital signs have worsened beyond anything seen before. The potential collapse of natural and socioeconomic systems, unbearable heat, and shortages of food and freshwater were cited as imminent threats. The report suggested that as many as 3 to 6 billion people could find themselves outside Earth's livable regions by the end of the century.

Let's not mince words. That means potentially billions of deaths.

Read this carefully to see how scientists use euphemisms: "confined beyond the liveable region" - why not just say "will be confined to an unliveable region". Then the great scientists' classic: climate change "causes" severe heat and limited food availability". Incorrect. That IS climate change.

The breakdown of climate/severe heat/mass death is one and the same thing. The causal agent is not fucking "climate change," it is the mass murder project of the world's elites. Every time a scientist or campaigner says "climate change" they create an artificial causal barrier between people who die and people who kill them. Given that humans are hopeless at understanding multiple causal links - the climate industry's key social role in using the term "climate change" is to protect the elites from being seen for what they are - the murderers of the human race.

Note on this: if you say A causes B then the vast majority of people will go - "Yes, A causes B;" but if you say A caused B which causes C, then hardly anyone will click that "A causes C;" A are the elites, B is climate change and, C is people dying

---

Source:
Article: Oct 24 AAAS News Release: "Climate report: 'Uncharted territory' imperils life on Earth"
Academic Paper: Oct 23 *BioScience* journal article: "The 2023 state of the climate report: Entering uncharted territory"

CLARE FARRELL

GAIL BRADBROOK

ROGER HALLAM

XR Turns 5 - Reflections on Telling the Truth

(Nov 19, 2023)

Telling the truth means being ready to accept when you have been led astray, got things wrong and so have to modify your views. Five years since the launch of Extinction Rebellion, this world we live in is changing so dangerously fast that it demands we revisit our assumptions and learn some painful lessons. It is now clear that 2023 is very likely to average more than 1.5 °C above a 1850-1900 baseline. Whilst emissions are still rising world wide. It is only through commitment to the truth that we might help humanity and wider life around as we enter a disturbing new era.

We got something wrong. We were misled. So, we misled you too. Aerosol pollution matters *decisively* to our global climate. There are other factors deserving of more serious attention such as forest cloud seeding and ocean health. Many factors were sidelined by scientists who were narrowly focusing on CO2. In addition, IPCC processes did not find an adequate way to address issues of extreme risk where data was deemed insufficient or where there was higher uncertainty *, such as aerosols, methane release from permafrost, and feedbacks from wildfires or droughts rendering sinks incapable of sustaining their role in the system. This misled other scientists, academics and activists including us.

Some of us have attempted over the years to responsibly communicate the extreme and cascading risks, and the severe consequences of not taking emergency action. Despite founding the movement on the precautionary principle we found ourselves being ground down. For years we were moderated, and moodsplained by experts from narrow disciplines who demanded we change our press releases, our lectures, and play down the reality and potential speed of catastrophic consequences. As we pass into the horrors of a 1.5 °C plus world, at least 10 years earlier than the worst official expectations, we realise we should have made a firmer stand. As we observe some top climatologists claiming we need to wait decades before accepting that the planet is 1.5 °C warmer, we also realise that silence about our disagreements is no longer an option for us, or the climate movement.

Richard Oelze, Expectation, 1935-1936, oil on canvas.

It is Better to Die With Honour Than To Die Like Sheep

(Jul 23, 2022)

Warsaw 1943 - London 2022

In 1943 word came down the line in the Jewish ghetto in Warsaw that people were being taken off to be gassed at Auschwitz. The dead weight of denial kept reality at bay. "Surely they will not have us die". "Surely we are of some use". "Evil of this depth cannot happen". The deniers were wrong. The Nazis wanted them dead but they told whatever lies were needed to keep everyone passive. Jewish police were co-opted to keep everyone in line.

When the ghetto numbers had fallen from around 300,000 to 60,000 the flip happened. They are going to kill us all. It was too late to live but not too late to resist. To die with honour rather than die like sheep. The Jews fought like hell by all accounts. Much longer than expected. But in the end, apart from a few individuals hiding underground, they were all dead.

The rumours had begun coming down the line in 1933. Some say from 1923 (Hitler's intention was clear enough in Mein Kampf). They grew during the 1930s. They screamed in 1939. Then it was too late.

This story used to be the greatest moral warning of European Culture. Never again. Never again. Now those who speak about our responsibilities are silenced. We cannot be allowed to hear about us allowing such evil to happen again.

The Left

Getting our Act together.

The Information Illusion

We don't need more information about how fucked we are. We know that. We need to know how to act together.

(Jun 20, 2024)

A week or two ago I went to Manchester to talk publicly about the fact that the human race is heading for extinction. I used what abilities I have to empower people to enter into resistance to stop this ultimate obscenity. I only gave one piece of information: recently the UN Climate Change Secretary, Simon Stiell, said there are only two years left to "save the world". Stiell added that he was not "being melodramatic" - meaning the world will not be saved without drastic action, meaning that if the world is not saved there will be "no world", meaning Earth will not be habitable, meaning billions of people will be killed. The issue here is not about the "information". Giving more information about this, in my view, would insult the audience. It would be as if your neighbour asked you to save their baby from a burning building and you got out your notepad and asked for more information about how flammable their bed is. If the house is on fire that is more than sufficient information to know what you have to do to - what a reasonably decent person like you has to do - get the fuck up and run over and help.

The problem is not the "information". "Information" is an abstraction. In the real world, it does not exist - it is a concept extracted from the act of communication which necessarily is embedded in a pre-existing matrix of meaning systems - an ever-interacting ecology of social, political, existential and spiritual orientations through which the "information" is made intelligible. Communication is saturated with what it is to be human - to be emotional - to feel.

To convey information that we are about to enter hell unless we do something drastic is a massively emotional event - triggering overwhelming distress, then repression, then dissonance in an audience. To get from there to action - i.e. to actually create change - the last thing people need is more "information".

What people need are three things.

First, they need to be helped to observe and then master how they feel so they can be how they really want to be. This can be done in several ways. They can be reminded that they are good people - or at least they would like to be good people. So what does that mean when faced with this reality? It means action. They need to be reminded that one day they will die and on their deathbed they will not be thinking about their bank balance, their status, or their security - they will be thinking about whether they loved the people they wanted to love, whether they made a difference in their lives. And this means that I have to act now. They need to be reminded that life owes them nothing - at any moment we can be subject to a life-changing accident, struck down with a terminal disease. We learn to accept these radical uncertainties: the necessity of accepting fate. Our fate in the present moment is to face this reality of social collapse - and the necessity to react to it in a way that conforms with what it means to be the people we find ourselves wanting to be. To get the fuck up and stop what is happening.

Here's Why the Greens Flopped in the EU Election

The Greens are down from 4th to 6th place in the EU elections in a time of climate collapse. Here are three reasons why:

(Jun 11, 2024)

1. They are not telling the truth for the sake of telling the truth. Last week a world expert suggested that we have a 40% chance of losing European civilisation in the 2030s because of the AMOC collapse. There needs to be a different tact - the situation cries out for raw emotion and real realism. No more being nice.
2. Greens support the status quo. Whatever they say otherwise, you can feel it. They are born out of the system (i.e., university-educated). Instead, they need to name the beast - capitalism - which is going to destroy itself and take us down with it. They need to say "Over our dead bodies will ordinary people pay for the emergency transition. It is the rich and powerful who got us into this total mess in the

first place and they should pay". And they should sound like they fucking mean it.

3. They should stop being a traditional political party and become a social movement that stands in elections. They should use community assemblies to choose policies and candidates including working class and people of colour communities. Then they come from the people and look like they come from the people. They should never go into alliance with social democrats and the other carbon legacy parties. And never should they be "responsible" and vote for budget cuts for their communities.

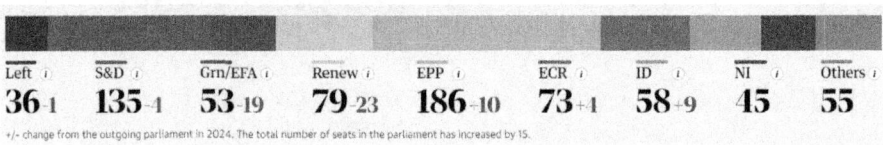

Left _i_	S&D _i_	Grn/EFA _i_	Renew _i_	EPP _i_	ECR _i_	ID _i_	NI _i_	Others _i_
36 -1	135 -4	53 -19	79 -23	186 -10	73 +1	58 -9	45	55

+/- change from the outgoing parliament in 2024. The total number of seats in the parliament has increased by 15.

European Election Results - from left wing to far right.

Of course, this won't happen - most Greens still want to have their cake and eat it. Maybe that was true in 2000, just possibly in 2010, but not a chance in hell in 2024. It's 1.6C - we will be over 2C in a decade, all hell is going to break loose, and the neoliberal regimes will collapse. Strangely, the Greens are the biggest climate deniers.

What we need is a whole new political system and revolutionaries to create it, who enjoy telling the truth and acting as if it's real because, at this point in history, they find expediency existentially repulsive (i.e. full of shit).

I know at least a few Greens out there know all this - so...?

Leading left-wing intellectuals, Paris, Copyright © David Seymour/Magnum Photos

The Intellectual Left Are Fucking Bollocks On "What to Do"

(Apr 4, 2024)

Guardian columnist Owen Jones and ex-city trader, Gary Stevenson's latest video, Why We're In This Mess is great on the no-brainers. For example, in a stagnant economy, if the rich pass a tipping point of wealth, they eat up the poor. Well, nothing particularly new there - just what Marx was saying. I played Monopoly every Sunday for years when I was a kid and, if you are reasonably clever, you soon work out that as soon as you tip into owning a certain number of properties the other players are exponentially done for.

All well and good. But then, in the last five minutes, they move to "We have to give hope" and well - what's the plan? Owen is "supporting" independents and Gary wants people to listen to him more: What I call the macro delusion and the information delusion. What they are saying is so dumb it makes me want to throw myself over a cliff. All they are doing is creating even more despair and thus passivity.

Gary, quite understandably, can't get his head around why, given he has earned shit loads of money, the government and academics won't listen to him. At the same time, both of them don't want to listen to people who know how to build political power and get taxes on the rich.

That's where I come in. My team have created the three biggest climate/social movements in the UK - XR, Insulate Britain, and Just Stop Oil. We developed a detailed integrated mobilisation model and replicated it in 10 Western countries - creating the biggest campaigns in 5 of them in two years. If you are a social scientist - i.e. you're familiar with complexity - you know that is pretty amazing. Just as you know Gary is cool for getting his predictions right several times in a row. Why? Because he has a different theory of how the world works - and so he can make lots of money.

So guess what - why have I been successful in building political power from literally £250 in the bank and 3 people in a room? Because I have a totally different theory of how the world works to all the NGOs that have 10s of millions and do fuck all.

Is anyone in the system interested in speaking with me? No. I have been almost completely ignored.

By "the system" I am talking about Owen and Gary, and all the rest of the intellectual Left who continue to be completely clueless on how to build political power. They haven't even got to page 2 on the basics. (All of which I've outlined in my podcast, Designing The Revolution)

It's Time For Prophetic Leadership — Interview

(Jan 19, 2024)

This interview was originally published in the Dutch Magazine, De Volkskrant.

Roger Hallam opens the door: 'Ah, the journalist person.'

His living room, in a small apartment in South London, is crammed with a piano, bookshelves, two bicycles, an armchair, and a sofa with duvets and blankets on it – apparently, someone is sleeping there. The curtains hang loosely on one side. Hallam offers water in a tea mug with brown residue. 'How long do you want to talk, by the way? I have an hour.'

An ankle monitor is strapped around one of his ankles because Hallam is under house arrest – he is not allowed to leave his house between midnight and seven in the morning. The arrangement has recently been slightly relaxed; initially, he couldn't leave his house after ten. Nowadays, he can visit his children, who live elsewhere in the United Kingdom. 'A political measure,' Hallam judges. Harassment by the authorities, just like the police raids he experiences every few months. 'They take my devices, interrogate me, all with the aim of putting me behind bars. It's state intimidation.'

Roger with his Ankle Monitor

A week before this interview, it happened again: the activist group Just Stop Oil posted a video of officers ransacking his living room. Hallam cheerfully looks into the camera and gives a thumbs-up before being taken away by the police himself. It has become business as usual. He estimates this to be his thirtieth arrest.

Hallam: 'These raids don't happen because I publicly call for roadblocks. If you've watched my speeches on YouTube, you know I never do that. I point to the science and the absolute necessity to resist. The law is politicised to create repression. You are now talking to me about civil disobedience. Are you also civilly disobedient? Under an authoritarian regime, probably. I'm afraid we are increasingly heading in that direction here in the United Kingdom.'

As a co-founder of Extinction Rebellion and Just Stop Oil, Hallam is one of the most influential environmental activists globally. You could consider him the mastermind behind the Dutch blockades of the A12 in The Hague – 27 days long and resulting in 9,000 arrests. Earlier this year, Hallam spoke several times via video link with Dutch activists, he says. Hallam was the one who said: you have to come back every day. 'They wanted to go to the A12 every month, they said. I said: that's great, but it makes no sense. You have to do it day after day. That's the only coherent strategy.'

Just Stop Oil, founded in 2022, gained global fame after two activists threw soup at Vincent van Gogh's Sunflowers (behind glass). But there have been numerous actions, from smearing buildings and glueing themselves to oil pipes to blocking roads. These weeks, 'slow marches' in London, where a group of activists walk slowly across the road, lead to a wave of arrests. Causing traffic disruptions makes Just Stop Oil a favourite target of right-wing politicians and tabloid newspapers, who speak of 'eco-idiots,' 'thugs,' and 'gangs' hindering ordinary citizens and costing millions in taxpayer money.

In recent years, the actions of environmental activists have been increasingly heavily punished. Two Just Stop Oil activists were sentenced to 2 years and 7 months and 3 years in prison, respectively, for climbing a bridge, causing a major traffic artery to be closed for hours. Ian Fry, the UN special rapporteur on climate change and human rights, said he was 'particularly concerned' about the sentences, the highest ever imposed in the UK for nonviolent protest.

Hallam himself spent four months in pretrial detention at the end of last year on suspicion of participating in a conspiracy to disrupt public order. A speech for Just Stop Oil activists had been secretly recorded by a journalist from the tabloid The Sun and leaked to the police, according to the tabloid itself: 'a victory for the people.' The newspaper then photographed the police raid on Hallam's apartment – a one-two, according to Hallam. He was not at home but was later arrested elsewhere.

Hallam: 'Make no mistake, it may seem to you that the Netherlands is friendlier, but we are just a few years ahead of you. It could go the same way for you.' In the 37 seats for the right-wing populist Dutch Party For Freedom, Hallam sees new evidence that a 'real left' story is missing. 'Neoliberal "left" breeds fascism,' Hallam wrote on social media after the Dutch election results because the left refuses to break with capitalism. 'We will only be saved by real left, which declares: "We will tax the rich, and they will pay for the carbon transition."'

In interviews, Hallam often clashes with British journalists, not only with those from the tabloids but also with those from the BBC. Hallam says, "Because they don't want to talk to me about the science. But if you, as a journalist, don't summarise the latest state of science, the public will think: this is just a strange radical saying crazy things to entertain us."

According to the annual greenhouse gas report from UNEP, the United Nations environmental organisation, released at the end of November, the current climate policy is far from sufficient to stay below a 2-degree Celsius increase. According to UN experts, the Earth is expected to warm by 2.5 to

2.9 degrees by the end of this century. Disasters, such as the melting of parts of the Greenland and Antarctic ice caps, are likely to occur, resulting in metres of sea-level rise. The monsoon in Africa could stop, the North Pole could lose its summer ice, and the glaciers in the Alps are likely to melt. According to UNEP, "ruthless mitigation and transformation to low carbon" are necessary, while greenhouse gas emissions rose by 1.2 percent last year, setting a new record.

Due to the climate crisis under the current climate policy, the World Bank estimated that 216 million people would have to relocate. UN chief António Guterres said at the end of September that "humanity has opened the gates of hell" and warned that we are heading towards a "dangerous and unstable world."

Hallam, trained as a sociologist, worked as an organic farmer in South Wales but stopped when prolonged rainfall led to the death of his crops. Hallam believes this is due to climate change. In 2017, he moved to London to pursue a Ph.D. at King's College on civil disobedience, a path he abandoned to focus full-time on his activism.

In 2018, he co-founded Extinction Rebellion with others but distanced himself from the group a year later due to conflicts over leadership and strategy. He wanted to fly drones near Heathrow to force the closure of the airport, but the action faced resistance from other activists due to safety risks. Hallam denies the risks, intending to launch the drones just within the restricted zone but at a safe distance from air traffic. He proceeded and was arrested. A lawsuit related to that action is ongoing.

Hallam also sparked anger among other activists when he described the Holocaust as "just another fuckery" in human history. Several XR chapters publicly distanced themselves from him, including in the Netherlands. Hallam also claims that global warming will lead to civil wars, which, in turn, will lead to, for example, group rapes, mass murder, and cannibalism. He describes these scenes in detail to journalists, stating, "They take your mother, lay her on the table, and rape her, then they grab a stick and gouge out your eyes. That is the reality of the destruction project we are facing."

In early 2022, Hallam founded Just Stop Oil, a movement explicitly focused on disrupting public order. The UK branch of XR, on the other hand, renounced civil disobedience, claiming it would be less effective, though stricter penalties might also play a role. Hallam sees it differently, stating that demonstrations, petitions, and lobbying won't move politics. He claims to know nothing about recent Just Stop Oil actions, deliberately keeping his distance as he feels he's being watched. He says, "I think at a distance about strategy."

Liberals

It's time to break free.

People Power Movement

A Primer on Reform and Revolution: An Open Letter to Green Party Members

Climate collapse has pushed us beyond the limits of reformism. It's time to follow the historical precedent and call for a revolution.

(Jul 3, 2024)

I have been thinking about the process of social change literally every week since I was 14. It is my life's obsession - it drives me. I have been organising people every week for the same period to make this world a better place. I have done years of award-winning research at King's College on the dynamics of political mobilisation. I have read the literature. Last year it was decided by the New Statesman that I was the 34th most influential progressive person in the UK - the only "environmentalist" above me was the untouchable David Attenborough. So, no pressure, but maybe you might take 5 minutes to read my thoughts on the Green Party.

My decision to write this letter was "triggered", to use that fashionable word, by the following from the Guardian in the run-up to the UK elections:

> "Asked if the Greens believed jailed climate protesters should be freed, Carla Denyer said it was not up to politicians to get involved in individual court cases."

In case you don't know, my good friend Phoebe, 22, is presently facing a long prison sentence for... yep walking on a road for 20 minutes to demand that her generation is not totally fucked over. And in case you are blissfully unaware of what it is like in the worst prisons in Europe, read about my experience in the English Gulag.

What I want to say is that reformist expediency in pre-revolutionary times, defined as doing "what works" in the short term rather than responding to what is objectively real, is not only morally obscene, but also strategically illiterate.

Let me start by illustrating this viewpoint with a little story.

I was an organic farmer for twenty years. It was like getting blood out of a stone. I asked people to spend an extra £5 to buy local veg that won't "destroy the planet" as the phrase goes. I loved the work but it made no money and I had a headache every weekend from the stress. I rang people three nights a week, every week - roughly 100,000 phone calls during those two decades - "to get peoples' orders" and try to persuade nice "green" middle-class people not to go back to Sainsbury's just because their broccoli had a bit of mould on it. Then Covid hit (the shock to the system) and everything changed. For the first time ever former customers were ringing up sheepishly asking to start their orders again. My lower self (which can get pretty low) was saying "fuck off go and starve" but, don't worry, I was very professional (having had a good Christian upbringing) and gracefully got them their veg.

Out in the polytunnels.

In historical sociology, the name of the game is to look at social patterns. One of the key patterns is the interaction between reform and revolution. These terms, formally speaking, have tight neutral definitions. Reform is to work within the system and try to change it. Revolution is to work outside the system to try and change the system itself. In reformist times (when objectively there is no material possibility of revolution) the reformists win out. In revolutionary times (when there is no material possibility of reform) the revolutionary wins out. In sociology, this is called structural determinism. Sometimes, not always, what is going to happen is determined. It cannot be avoided. It's bleeding obvious to anyone looking at the stats.

Let's look at an example.

When Lenin arrived in Russia before the revolution, he looked at the stats and knew there was not a chance in hell of the Russian army beating the Germans in World War One. They had few weapons, no supplies and morale was non-existent. He decided that the revolutionary Bolshevik programme would be to stop the war with the Kaiser whatever the cost. All the reformists thought he was a complete idiot and his party was immediately looked upon as irrelevant. Supporting "Mother Russia" in the war was ideologically non-negotiable. And so the reformists, to cut a long story short, decided on an offensive to prove that belief trumps reality. Things looked good for the first day and then the whole front collapsed as the Germans advanced another few hundred miles into "Mother" Russia. Reality trumped belief. The Russian masses turned towards Lenin for peace, not least because he had been right all along, and the rest is history.

It was structurally determined that the Russians would lose the war and that this would create a revolution. The same can be said of the situation before the French Revolution. It was structurally determined that the old regime would collapse because of the massive increase in state debt. Then the finance minister finally dared to tell the King that the state would totally run out of money in six weeks.

Here is point: Expediency does not work in a pre-revolutionary period where the wider context determines the end of the regime. Though those policies, be they war with the Germans or increases in French state debt, may help maintain the status quo in the short term, they will ultimately lead to its collapse.

Let's go back to organic farming. I was seen as dumb because it would be so much easier to screw the environment and just become a normal "commercial" grower. But Covid was our first system shock: a little World War One, a little "running out of money", and then the tables turned. Everyone wanted my veg.

When Green Party leaders decide to publically shit on imprisoned young people in resistance to objective climate collapse whilst they play the "sensible adult" role of getting into a broken, reformist parliament, they are taking the expedient route.

Obviously, this is the immoral path but the logic of reformism is actually to have revolutionaries imprisoned. The Kantian imperative to not use people for political means is ignored. This is what it means to go after power. After all, you can't make an omelette without breaking a few eggs.

In reality, this is strategically super dumb because neo-liberal regimes are going to collapse with billions of refugees and mass food shortages. These are objective conditions that trigger a revolution in the structural determinism. Just as it was obvious that Russia was going to lose against the German army and the French state was going to run out of money - so too, at 2C, it's obvious that Western political regimes are going to collapse. The fact that this is not seen as obvious is because of the power of ideology over facts. In Russia the patriotic ideology that "mother Russia" could not be defeated blinded people to the country's severe losses. The neo-liberal progressive, humanist ideology of the Green party blinds it to the facts in 2024. Their belief is that the present regime of material prosperity based upon the rape of nature can keep going - that you can go green and keep capitalism.

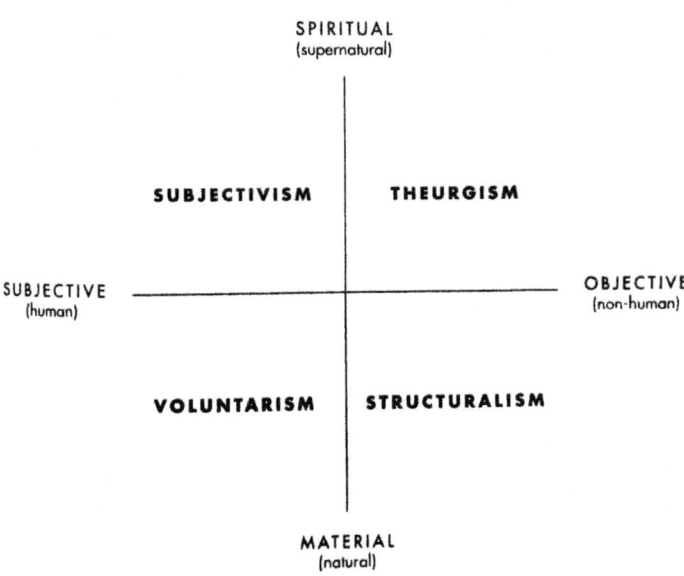

UNIFIED THEORY OF REVOLUTION

A unified theory of revolution is the quaternity of voluntarism, structuralism, subjectivism and theurgism.

Micah White's Unified Theory of Revolution from his book The End of Protest. The diagram illustrates how a variety of factors like structuralism impact revolutions. One example given is a food deprivation index that was used to predict the Arab Spring uprisings.

In contrast, are the facts made clear by the UN, the equivalent of generals sending messages from the frontl. Recently they said we have two years to save the world. They are not "being melodramatic". Failure will "decimate economies", not unlike the fiscal crisis before the French Revolution. This is UN "General speak" for hundreds of millions of people starving to death. What exactly do you think NOT "saving the world" will look like?

Get it? Leaving aside the small matter of moral corruption - the Green Party is backing the wrong horse!

In 5-10 years from now when Covid-type climate shocks come thick and fast, the UK government will collapse and there will be revolutionary change. That means a change in the regime and the way of making national decisions: mass social disruption which changes the constitution. Either the regime will collapse into some form of fascism or we will have a democratic revolution that comes from outside the system - led by people like Phoebe. People who were "irrelevant" and never "clever" enough to get into power are suddenly "thrust into greatness", to quote Shakespeare. Read the history of revolutions - it has happened a hundred times before. One minute some nerdy intellectual is cleaning windows, weeding carrots, sitting in a cell. The next minute she is running the state. "Out of nowhere" as they say. It's called history. But the core ideology of our time is that history ended in 1989. Green Party leaders feel they don't need to read history and so don't have to think about revolution.

The Guardian Facilitates Genocide

Let me show you why this newspaper has lost its credibility as independent journalism.

(Apr 20, 2024)

An article this month says:

> The Environment Agency released a report last week that predicts a growing shortfall of water in coming years, leading to a deficit of almost 5bn litres of water a day by 2050.

1. This implies that the "prediction" is a 5 billion shortfall (a third less than the present supply) meaning that is what is going to happen. But this is the average prediction - the top of the normal distribution curve. Obviously, there are the outlier long tail possibilities that it could be say 2 billion or say 12 billion. The average is not relevant - what is relevant is the catastrophic chance of a 12 billion shortfall - i.e. the total collapse of water

for agriculture and supply to urban areas (i.e. evacuation from cities).

2. The statement implies it will be 5 billion a day each year, but that just means an average year. Again, say 1 year in 4 there will be a 12 billion shortfall (see consequences above). And mathematically that means one year in 16 (.25x.25) this will happen 2 years in a row and will therefore be 100 or 1000 times more destructive - i.e. permanent migration from the UK and/ or social collapse (think about the Gaza crowds wanting to get food).

3. Last but not at all least - there is the old chestnut that this is a closed sub-system analysis. What else will be happening in the world by 2050 in a post-2C world? Well, according to other *Guardian* articles there is a 50% chance AMOC will collapse, leading to a 10-30C collapse in winter temperatures. Oh, and loss of control of wetland methane release, taking temperatures over 3C, runaway sea level rise (Greenland ice collapse), and then 1 billion people on the move because of wet bulb effect mass deaths in the subtropics.

So...

Independent journalism would insist that each article on the future would look at the whole system and add in the new elements of that day's news. Anything less is not worth the paper it is written on.

They would write, for instance:

"We have news today that the average collapse in water supplies will be 5 billion litres a day - a third of present supply, by 2050. We have to add that into the *Guardian*'s running all-system analysis of the future - 10 million extra migrants coming into the country and a 50-50 chance of mass migration out of the country due to AMOC collapse. We need to add in the inevitable collapse of the world economy at some average point in the 2030s.

Needless to say, given the interlocking facts, the *Guardian*'s independent position is further strengthened that revolutions are now inevitable, and they are justified and necessary for liberal civilisation's survival. Please sign up to join the *Guardian*'s staff team's unlimited hunger strike to demand a citizen assembly-led government, happening on June 1st."

A final point - it's interesting isn't it that when the *Guardian* does an article on individual suicide there is always a link to support services and the end - all right and good. But when it comes to the mass killing of billions of black and brown people, followed by the mass suicide of our whole civilisation - well it is "just another day" as the slave-supporting *Guardian* of the nineteenth century would have put it.

Independent? - Nope - complicit in genocide? Yes

The Schizophrenia Machine

The annoying thing about liberals is that they hate unpleasantness when it happens to them but they spend their lives not thinking straight about the 10 million times more unpleasantness they create for other people.

The *Guardian* is a massive schizophrenia machine. Over in the "climate department" we are told on a daily basis that the human race is heading for extinction - or effective extinction if you want to be nerdy about it. While over in the politics department it's liberal business as usual. This is not smart.

As columnist Marina Hyde says.

"I can see Ms Lunnon thinks that the only context for anything anyone says or does, ever, is that we are about to become extinct, and that trumps absolutely everything."

Let's be polite and suggest this sentence is not very well thought through. Obviously, if we were about to become extinct then that *would* trump absolutely everything. In World War Two, 3% of humanity died and the war definitely did trump everything - so presumably if 100% of humanity is about to die then...well.

The point, however, is not about basic logic. It is that Marina, along with the rest of the liberal elite, simply does not want to believe what they do not like.

How does this denial happen? I offer the following reasons.

1. *Guardian* writers think climate is an issue. It is not. By definition it is everything, in the sense that air and water are everything - no water you die - no climate you have no food, you die.

2. *Guardian* writers think climate is an event: "When we get to 2C..." - implies we stop there. No. 2C locks in 3C and so on. It does not stop.

3. *Guardian* writers think the climate is about the physical world, which they are dumb enough to think does not interact with the social world. If you, for instance, suggest that climate will create mass rape, all hell breaks loose. But to deny this is like saying a meteorite landing in New York will not kill people.

They think then that there is only one domino. There is in fact a series of them. When you knock down one, you will knock down the rest of the row. This is basic physics. The tipping points in the geophysical system trigger each other, which then re-amplify each other - it's ONE SYSTEM.

This is how you lock in human extinction. And everyone with minimal analytical skills knows that human extinction is now the main scenario.

Of course, saying this is not going to get Marina to understand what is going on, because people who are embedded in a self destroying elite system can only very rarely get their heads out of it.

This is why these people have a complete lack of understanding about why people think it's okay to hassle those who administer the greatest act of injustice in human history. They don't realise that the old liberal system is done - it's totally fucked. The future is fascism or some form of Citizen Assembly-led government. Either way politicians have no future. Sorry.

If you stand by and allow the world to experience something on the scale of 30 world wars, you really don't get to complain about a bit of vigorous "protest".

> **Note:** The reaction to this will be all about the vigorous protest and nothing about 30 world wars - the joy of privileged delusion.

Labour's U-Turn on Citizens' Assemblies Shows They Fear the People

(Feb 20, 2024)

The Labour Party's 24-hour U-turn on introducing Citizens› Assemblies is testament to the radical idea of actually letting ordinary people control their own future. When you realise that, it's no wonder Keir Starmer fears them. He should.

Citizens' Assemblies are our last hope for rational governance in our age of climactic collapse. But the 'powers that soon will be' were keen to co-opt that. They only want to use them to patch up the fraying legitimacy of the neoliberal regime. That's kind of what the Labour Party does.

But once the spectre of real democracy was raised, the Party dropped the idea like a hot rock. Just like when Labour responded to Corbynmania by desperately trying to stop ordinary people joining the Party.

Citizens' Assemblies have been the wet dream of the liberal middle classes for a while now. It's like the last cake they can throw out the window of their ivory towers before the pitchforks come out. It won't work.

Sketch from Adolf Eichmann's Trial

A Small Matter of Treason: Starmer and The "Climate"

(Feb 20, 2024)

For the past five years I have read ten to twenty articles and scientific papers every week on the "climate", i.e the coming holocaust enacted by the global elites. I recently received one that predicted a 40% reduction in global GDP by later this century. This builds on two other key papers which predict over one billion migrants as we shoot past 2C in roughly the next 15 years and the resultant billion deaths by the end of the century. So if this is not a holocaust, what exactly do you think these predictions mean? Have the decency to be honest with yourself please.

I used to be a social science researcher at King's College, London. The name of the game in that trade is to look at everything in context. So I will provide some context. There is overwhelming evidence that the climate science industry is structurally underestimating the realities we face. "Worst than expected" is the standard phrase of just about every article as new stats are published. A few years ago we were going to pass 1.5C around 2050 - now it is already happening. I remember reading reports that the Arctic will melt in the summer around 2100. Papers now predict 2035, if not before. AMOC - the ocean current that stops the 60 million people on these islands from starving to death - was going to collapse at some point next century. Now a recent paper tells us the odds are it will collapse by 2050. If you have not been paying attention, this will create a collapse of temperatures overnight of 3-8C across Europe. So don't be surprised if it happens before your pension comes due.

I'm like you. I don't like to believe things are true if they conflict with my baseline beliefs - like "we will muddle through". But then it becomes more difficult when it actually comes true. Scientists have been telling us privately and then publicly for years that staying under 1.5C was bollocks - and now here we are. For two decades or more the best kept secret of the climate space has been that aerosols (pollution from burning fossil fuel emissions) have been holding down temperatures by .5C-1C. As we passed 1.8C last September the pretence started to collapse as scientists raged about each other on a dark corner of Twitter. It's the start of the exposure of the world's biggest cover up. That they knew we were fucked a decade or more ago. Not that the media is interested. Everyone is still in on the pretence, it seems.

Including Mr Starmer, as he decides to go back on investing a few billion in saving what is left to save. The phrase that rings through my head as the next generation takes him to trial for genocide at some point in the 2030s is "you knew and you did nothing". No doubt Mr Starmer's top legal team will point out, echoing Adolf Eichmann, that he did not actually do the killing. But it won't wash any more in 2033 than in 1963. It will be a political trial. Starmer, like Eichman, will be guilty before he enters the courtroom. They hanged Eichmann. No one gets to kill Jews, and no one gets to send the next generation to their death either - without consequences.

The Gleaners (1857) by Jean-François Millet

On The Farmers Protests

(Feb 2 , 2024)

I was an organic grower for twenty years - I would work 60 hours a week, bring up my kids, and the only highlight was going to town on a Saturday. Every weekend I would have a crippling headache from the stress.

Why? Because people who do the real work, the essential work, get paid shit wages while the rich pay fuck all taxes. Everyone is sick to death with the humiliation. On top of that, the elites are collapsing our climate, destroying everything in its wake. Normal people don't want a lot - just enough and the dignity of not being treated as idiots.

Unless new progressive social movements arise that make a clear break with the suicidal bollocks of the neoliberals - Starmer, Macron, Scholz - fascists will be in power within the decade.

The new offer is this: "Yes, the country will be less "rich," but at least we can have equality, dignity and community - and that is what counts."

We face mass migration and the collapse of world trade - insurance markets, coastal property prices - as climate breakdown destroys the global economy. Combine that with fascism and you have the biggest shit show in human history.

The situation could not be more serious. Wake up and act.

The British Establishment is A Cult: My BBC Interview

(November 16, 2023)

Like other interviews I've done with members of the political class, this one again shows their psychotic inability to see that the real world trumps the political world. The contemporary British Establishment is a cult that believes in two extreme worldviews:

1. **Vulgar Utilitarianism:** the complete inability to see the value in doing anything because it is good in itself. Everything is a function of the question "does it work?" – which practically results in a chronic short- termism, and the notion that ends always justify the means (e.g., it's okay to lie if it will take you into "power").

2. **Absolute Post-modernism:** the complete inability to see that some things are objectively real, such as the laws of physics. Everything is seen as a "point of view" and a "belief," apart from

the belief that everything is subjective which is insisted upon with rigid dogmatism.

The reason the political class is taking us to civilisational collapse is because it cannot think straight. It has lost its moral compass and any analytical intelligence. It exists in its own closed world.

Worst of all, it cannot feel emotion – the entry point for a change in one's worldview. When did you see a political or media figure cry about the betrayal of our youth, our country, our world?

The political class is DEATH itself – there can be no compromise with it. It is never going to save us in the time we have left.

We can only save ourselves through a political revolution which puts ordinary people in power through assemblies – not because "it will work," but because it is the right thing to do.

Louise Bourgeois, *Fears*, 1992. Wood and iron.

Visioning Extinction Rebellion

(February 4, 2021)

Extinction Rebellion was set with two purposes or visions – to speak the truth and to act as it is real. This sounds straightforward but it is in fact, at the present time, the most difficult thing in the world. That is why we are heading for extinction. Because we put our individual lives before the collective life of our species – before Life itself. And we wallow in self pity rather than stepping up.

The original vision of XR was not of a campaign or plan – but new way of being in action. A taking up of courage – a realising there is something more important than our fears. We all fail at this every day when we pretend we cannot do more. We hide from the world. We hide together and share stories that blind us to our cowardice. We are lost, but worse we resist being found. Because we do not understand that redemption only comes through suffering and the only honourable life is to move into that suffering in an act of faith that there will be another side to come out of, into a state of grace.

XR created a vision and set off to fail at it, and it has failed. And this is the first and necessary step towards success. The question now is whether it is willing to learn – to take the risk of destroying itself in order to reconstitute itself to produce a second iteration – a second chance of success – another second possibility of humiliation.

Only when we understand the notions of faith and grace will we find ourselves in our lostness. Only when we admit the utter destitution of our souls at this time of utter annihilation will we begin a journey we can be proud of, regardless of the outcome.

Two years after XR started I found myself last week in a zoom call with just four people from a call out for a third of the country, for people willing to undertake the minuscule commitment of going into a British prison as a result of their resistance to the most absolute evil any generation is ever likely to face. In the utter sense of failure in being unable to persuade this movement to commit to even a small fraction of the commitment our forebears expended to produce all our comforts and securities, I find a sense of grace.

My whole life has been an utter failure. An utter inability to stop the cruelties and injustices that people engage in. But long ago I learnt that beauty lies in trying, not succeeding.

The original vision of XR then was not so much to tell the truth and act as if it is real – but to fail at such a vision, and allow ourselves to enter in a state of grace about this failure.

And then, and only then, will we lay the foundations for the miracle to happen, that will save us – as it is written in the great stories of our ancestors.

A Future Vision

My future vision of XR is to let go of ourselves. And let go of each other. And through this we give up on our desire for agency. After all it is the will to power that got us here in this terrible situation in the first place.

We will only save ourselves when we put ourselves to one side.

We will not save ourselves though "caring" for ourselves but by abandoning ourselves, because when we cling onto ourselves we open ourselves up for needless suffering – particularly because we stand to be annihilated in the coming decades. And more deeply and perennially, because we are only here in this world for a short time.

My vision is that, only when we give up, will there rise up within us the sense of service and the power of fearlessness that will bring down the dark empire we face – the pervasive deathliness of everything this system of collective being has set up for us.

Only then will we gladly be able to give up all we need to give up in order to effectively rebel.

Specifically: To stay on the streets till we are banged up in prison for weeks and months. And then come out and gladly do it again. Because we literally have nothing to lose. What we have to lose is of no importance any more.

Maybe my vision here is that some people get a sense of what I'm talking about, and they will become the leaders of this movement – the prophets if you like – who will show the way.

Show people the true Spirit of Rebellion.

Rebellion is not a brand or a campaign or a frame or fashion – it is a way of being in action. It is a fusion of calmness with absolute transgression. A calmness in absolute rage. A stillness that comes from freely entering into the wildness of Life's desire for itself.

There's lots to organise of course. In the day I'm a thorough going materialist – but at night my soul sings to the cosmos and the cosmos sings back, and I wake with madness in my veins.

The Sublime Madness Chris Hedges speaks of. The madness that will bring down this Empire of Death.

That's my vision of the future. It's basically an invitation to enter the eternal adventure of the human spirit. To what it means to truly live a life. To welcome everyone home.

ROGER
HALLAM

The Liberal Class is Complicit in Mass Murder

(September 14, 2020)

Centuries ago, Edmund Burke wrote that the triumph of evil required only that "good men to do nothing." The notion of "evil" has been banished from the supposedly sophisticated discourse of the "woke" liberal classes. But for 99% of human history, evil was very much a material reality, namely the grotesque arbitrary power of the rich to rape, starve and murder. The modern evil is the plan by the corporate elite and their political administrators to willingly, in the full knowledge of the science, engage in putting greenhouse gases into the air to the point of locking in what is euphemistically called social collapse, namely that old trinity of rape, starvation and murder, on the scale of billions of people.

A recent paper outlined that one billion people would be forced to migrate (that means three things, see above) at 2C average global temperature rise*. And if you haven't been paying attention, that is now effectively locked in within 20 years, give or take a decade or so, unless there is emergency action to slash carbon emissions. This is not alarmism, it's the real world. Evil is always with us. There are always people committing monstrous crimes in order to express their disdain for life through a pathological desire for power.

The "good people" are supposed to be the liberal class, the professionals and administrators, the educated strata of society, who have the agency, the power and ability to stand up for civilisation.

So,where are they at this most critical point in the history of humanity when over the next decade, we are set to lock in the most despicable scheme to destroy a thousand years of social progress and condemn the next generation into a hell hole of endless and indescribable suffering? They are nowhere to be seen. As co-founder of Extinction Rebellion, I have spent two years asking nicely, waiting patiently for the NGOs, political parties, trade unions, churches, and professions to step up from their business-as-usual mode of organising emailing, lobbying and occasional tame marches.

What is now needed is bleeding obvious: mass participation in civil disobedience, escalating to a point of bringing the carbon death project to a halt. After all, this would certainly be happening if those already dying were in the home counties rather than in Africa. Black lives do not matter to those liberal classes as if we ever needed reminding. The Extinction Rebellion plan as I conceived it was that by this point, the educated classes would have done the maths and worked out we are going to hell unless we rebel. Our ask to civil society organisations was not complicated: leaders would go on a tour to speak to the nation, emails would be sent to the millions on their databases, informing them of how to engage in civil disobedience. Their vast organisational infrastructures would move into gear to coordinate the simple act of sitting in the roads.

What did we get? Greenpeace refused to send an email. Christian Aid was going to think about it. Charities were going to put It on the agenda for board meetings. The supposed "radical" mayors of Hackney and Newcastle rubbed their chins. The result: nothing. One year on from Parliament acknowledging the reality of the biggest genocide project of all time, the "climate emergency", all we have is inertia, procrastination, and outright opposition to anything resembling civil resistance. The Labour Party reverting to type plans to drop its commitment to zero emissions by 2030. The Green Party, bless them, only discovered the climate emergency once Extinction Rebellion appeared and since then have done nothing other than talk about it. Dozens of councils have declared an emergency and then promptly put their motions in a drawer and carried on as normal. Why? Because the liberal class have no guts.

The days of standing up to tyranny have long faded. The life-and-death struggle against Hitler and fascism is consigned to the history books. Today's liberal classes believe only in one thing: maintaining their privilege. Their one priority is power. The number one rule is: preserve our careers, our institutions at all cost. The historical rule number one of fighting evil is the willingness to lose your career and to risk the closing down of your institution. The prospect of death and destruction is lost in a postmodernist haze. Leadership has decayed into sitting behind a desk, following public relations protocols (otherwise known as lying). Leading from the front, the first to go to prison Martin Luther King-style died with the passing of the World War II generation.

The game is up. The old alliance with the liberal classes is dead. New forms of revolutionary initiative and leadership are rising up. Members of the new political party Burning Pink have thrown paint at the doors of the NGOs and political parties calling for open dialogue and public debate. The response, true to form, has been a lethal and deafening silence. We are now in prison from where I write this article after a Green Party member recorded a zoom call and passed it to the police. We have not been let out for exercise for the first five days. We have no kettle, no pillows, no visits. But we don't give a shit. We are doing something about Evil.

Shortly after the April Rebellion in 2019, I met John Sauven, chief executive of Greenpeace. He congratulated me on XR's success. He said he would have got arrested but he didn't because he had a cold coming on. The next generation have their annihilation coming on. They don't want to be offered cake!

Roger Hallam, Pentonville Prison, London. Received on 11 of September 2020.

Conservatives

Either we work together or we die together.

Hope Against Hope:
Recruiting Conservatives

(Jan 10, 2024)

(This article was recently rejected for publication by Unherd, a social conservative media group focused on preserving traditional values of culture and power. It is dedicated to Chris Skidmore, who resigned this week in protest against new oil and gas licences, making him the only true conservative in the House of Commons).

The second most depressing moment in recent years was receiving an interview invitation from a journalist at Unherd (1). He merrily said to me before the recorded conversation, "let's not talk about science". Which is like talking about Attila the Hun without mentioning the killing. The fact that a supposedly serious conservative political media organisation has no interest in "what is actually going on" sums up the mess of our times.

In many ways the interview went well as it opened a few doors. Firstly, I surprised them by suggesting the end of civilization might actually upset conservatives. Secondly, it led to a few nice chats with a Tory MP and a bigger chat with a bunch of conservatives at the Houses of Parliament. Predictably, when I mentioned that the bullet of deadly carbon emissions had already left the gun but not yet entered the head, I was told that being negative was bad for getting votes.

The Tory MP told me afterwards I was the only true conservative in the room, and for a moment I thought I might have finally found the Churchill of my dreams. But as soon as things heated up, he ignored me. He cut himself off when Insulate Britain blocked the M25 for the 'revolutionary' demand of better housing insulation. He has since refused to answer my occasional emails along the lines of - "Hi, remember me? The science doesn't go away just because you are playing hard to get" (2).

System Change: Chatting with Conservative MPs

(October 13, 2023)

Can we see beyond our political divides towards a path to system change? That was the question posed to Roger and Zac Goldsmith MP by comedian Heydon Prowse at the Planet Local Summit. It was a dialogue that, despite political differences, shared many goals. Both wanted to challenge concentrated power and embrace the British tradition of open debate. They varied, however, on the roadmaps forward.

The following is a transcript summary of Roger's speech.

I want to thank Zac for sharing a platform with me. No one in the Labour party would even dream of doing that.

I'd like to start by quoting from a recent peer-reviewed paper:

> "If warming reaches or exceeds 2° centigrade this century mainly richer humans will be responsible for the killing (**Note:** not "the death"), the killing of roughly 1 billion mainly poorer humans"

The first thing we need to understand about the elites, with no disrespect to Zac, is they are mainly psychotic. I think that's the correct medical word for people who are fully aware that we are going into an era of mass death and pretend that we're still in the 1950s.

I'm a revolutionary. I foresee that the present political regime in other words, the British establishment, will not exist in the near term. For the record, I want to say that I don't actually think revolutions are a great idea. In a perfect world we could all get together as nice sensible people in Bristol and sort everything out. But we know that we don't live in that world. The elites will not change their trajectory from taking us to mass death and they will collapse under the weight of that contradiction. Or in other words, under the contradictions of capitalism.

The next ten years will be exciting in a certain Nietzschen sense because there's been a lot of emotional repression in the green movement and it has increased exponentially over the past decade. People have been desperately holding on to a reformist pathway to change. Now, increasingly they are realising that it's simply not going to happen. There's going to be a big rupture (That's an analytical point, it's not an ideological point).

The default position is a move towards extreme right-wing populist as happened in the 1930s, however what I'd like to say is that's not deterministic. Historically, it is possible for movements to resist that default descent into fascism when major crises happen. What's coming down the road presents us with a massive opportunity to create a pro-social world but we're going to have to work at it a lot.

So what is system change? By definition it's about looking at the whole system and how systems change and transform themselves. I'd like to address that because I think both Zac and Heydon have made various assumptions about how revolutionary change happens which are not sociologically robust. I want to make two corrections.

The first correction is that historically revolutions are arguably initiated by conservatives. That's because the system itself has become revolutionary in the sense that it has become destructive to the core connectivities of society. It's destroying society through its own power. For instance, in the 19th century when there were Revolutions in France in 1830 and 1848 a lot of the people wanted to go back "to a traditional small scale agrarian society". They were having a revolution against the power of capital. This is one of the reasons I have no problem in sharing a platform with Zac because there's elements of conservatism which are radically opposed to the power of capital. Their revolution is against the revolutionary violence of international capital. So, there's complexity and ambiguity in the word Revolution.

The second point is that a revolution does not tear down the state. The revolution reconstructs the state from a regime that is destroying the state. If we return to my opening quote, you shouldn't have much of a problem imagining how states are going to destroy themselves in the context of 1,000 million people dying. Afterall, that's the equivalent of 20 World Wars in the next two generations. So what we need to focus on is a project that is going to bring together traditional conservatism with traditional progressivism against the real enemy which is international Capital.

International capital is obviously a simplification but what I would suggest is the creation of a new political force. One that does not want to stand in elections to recreate the party system but wants to remove the party system and replace it with citizens assemblies. That's the international agenda amongst people that are thinking clearly.

If we don't get organised we're going to have fascism. Fascism is massively violent and in the modern context it will lead to human extinction. It hasn't got any universal rationality so it's easy to joke, as you do Heydon, that popular democracy is messy. Yes, It's filled with ordinary people and their different quirks. But as Democrats that's all we've got. We've only got the people. What we have to do is bring people together with our best social processes and enable them to feel their power.

We've had 40 years of the neoliberal lie that people can't organise themselves. That they're not interested and that we should leave it to the NGOs and the political parties to sort out this mess. We know that's led to the point where only 8% of the people in this country believe in the present constitutional Arrangements. This country is as mad as hell if you haven't noticed. On the surface we're being polite and English about it but underneath we're all mad as hell. We've just totally been fucked over year after year. Just in the last fortnight we've suffered the astronomical humiliation of a prime minister committing himself to genocide, to increasing carbon emissions, on our watch. So yes it is going to be really difficult. It's going to be hellishly difficult. We all have to grow up a bit and realise that if only three people turn up to your first Bristol people's assembly, you just stride right ahead because you haven't got any other choice. When we started Extinction Rebellion, there were two of us doing mobilisation. My friend Robin had the west side of Britain and I had the East Side. That's how it started. For the first meeting I rang my mate up about giving a talk in Nottingham. He said "Great, it's a cool place and we want to rebel. I'll get all my friends." The event turned out to be him and one friend.

Don't think there was anything glamorous about starting Extinction Rebellion. Everyone thought it was a twatty idea for at least three months and then 6 months later it had 200,000 people involved. That's how history works. Everyone thinks you're a dick and then suddenly you're a hero.

We're on the edge of hell and we all need to get a grip. We all need to work together to create assemblies around this country to revitalise Democracy. Those assemblies need to aggregate into bigger assemblies and they need to show their power through local direct action and manifestos.

By standing as Independents in elections we can constructively prepare for the revolutionary episode that's coming down the line. At some point the system is going to destroy itself for the reasons Zac pointed out - vested interests are destroying our state and society. We need to be prepared for it. When the moment comes, we must stop the forces of fascism from destroying everything we believe in.

To be a bit patriotic, this conversation between me and a Tory MP is a Glorious thing about British culture. Look, I haven't thrown water over him or got my walking stick out for any blows. We can have a reasonable discussion about our future. Our country has got a very problematic history but at the same time we need to look on the bright side. There's a lot of common sense and that's what we need. A common sense Revolution where we say "we're not putting up with all this bollocks".

You've got my email address so contact me if you want to get involved in assemblies. We're all going to work together to build something new. I'm working with a bunch of people, even relatively rich people. Obviously I'll be asking Zac for the odd million or two after this meeting because he's a patriotic guy and wants to see the country survive.

The point is we need to build another mass movement like XR which is well organised with a Central Committee. Sorry for those of you who don't know like Central Committees but that's the way you get things on the go. That's why Just Stop Oil has a 90% name recognition. It's properly organised. We're going to go out and become a voice in the public sphere of this country because it's desperately needed. The Labour Party is fucked isn't it. Let's be honest with ourselves that it's gone into the depths of Mordor and that hurts. We need a new story in this country of common mobilisation and organisation from the bottom up. The green movement is really good at making wish lists but we need to concentrate on the aggregation of political power. That's the chicken that lays the golden egg. That's what you saw with Extinction Rebellion. That's what you're seeing at the moment with Just Stop Oil. The reason why you know about us is because we're powerful.

We're powerful because we disrupt. Nothing's going to happen without disruption. The other side of the strategy is popular mobilisation in local communities. It's the Synergy between the two that provides the possibility of collective redemption.

Civil Resistance:
Between Protest and Just War

(Nov 4, 2022)

"If you believe that you are literally saving the world, that billions will die if you do not get your way, that an apocalypse is imminent and that your targets and opponents are what stands in the way of your efforts to save everything alive, then surely anything is justified to secure your ends?"

Mark Wallace, Chief Executive of the website Conservative Home, 1st November.

There are three types of action. Undisruptive actions, actions which are disruptive without justification, and disruptive actions which have justification. Claiming that all disruption is unjustified is dishonest – bad faith. No one believes that, and people who claim to believe it are lying to us, or worse, lying to themselves.

People support protest as it is non-disruptive - it simply expresses an opinion. It does not aim to coerce through disruption. People – at least most people – support just wars. Violent disruption is justified as a response to violent aggression. It is self-defence. It is justified for people to be killed in support of that self-defence.

It is therefore undeniably inconsistent to say, without exception, that civil resistance is wrong. Civil resistance is a non-violent form of war: "war without violence" as Gandhi's campaigns were called. In civil resistance – nonviolent disruption - people are not directly subjected to violence, but harm will come to bystanders as it does in violent just wars. Gandhi initiated the Quit India campaign in 1942 with the full knowledge it would lead to thousands of deaths. Martin Luther King led the campaigns in Birmingham and Selma with the full knowledge of the violence that would be provoked, indeed that is why these cities were chosen. This form of disruption is justified because it was responding to violent aggression – attacks on fundamental human rights.

Saying you support just wars but never civil resistance is incoherent – it makes no sense.

The question then is not whether civil resistance is good or bad. That is a silly question. The question is when is this type of disruption justified by its context? Mark Wallace, Director of the Tory website Conservative Home is intelligent and brave enough to make this clear. If we are indeed facing an apocalypse "then surely anything is justified to secure your ends."

We are facing an apocalypse. And so civil resistance is justified.

What resisters and conservatives have in common, in opposition to the liberal ideology of everlasting progress, is their acceptance that humans are fully capable of extreme violent aggression – evil is not a taboo word. The difference presently between resistors and conservatives is that the latter are still lying to themselves about what is going on – about the evil that is happening. They are betraying their long-held self-understanding that they "see the world as it is" rather than how people would like to see it. They have been infected with the liberal dreamworld notion that things will always be nice and fine. Things are not going to be nice and fine, and they need to be straight with us and themselves about this.

Conservatives believe in personal responsibility. None of this "I was constrained by factors outside my control" nonsense. You have free will. When you see evil, you have a responsibility to act against it. And if you don't you will be prosecuted for your crime of inaction. Your complicity. Your treason.

In the years to come, it will be conservatives that will be taking climate criminals to court. They will say "You knew and you failed to act. You are guilty and you will be punished".Sir David King, the former chief scientific advisor to the UK government, recently stated that if methane comes out of the melting permafrost due to continued carbon emissions, temperatures will rise by 4-8C in two decades. If this happens it will certainly lead to billions of deaths. It is self-evidently the greatest of all crimes to allow this to happen. This is the real world conservatives are supposed to like to talk about.

The great radical Saul Alinsky said – if you don't like the means it's because you don't believe enough in the ends. The end is to stop climate breakdown – billions of deaths and that overwhelming justifies the means of civil resistance – nonviolent disruption and coercion, a war without violence - to stop making this greatest of all mass murder projects from becoming a reality.

Justifying civil resistance can be complicated. Justifying civil resistance at this stage in time is not. In fact, nothing could be less complicated.

The only question is how long it takes for Mark Wallace to stop lying to himself. Because this country and this civilisation need him and fellow conservatives to get a grip. Time is running out.

News Commentary

My take on the headlines.

Threads movie poster.

Labour's Election Policy? To Shatter The Young's Bones For Eternity

(May 23, 2024)

Keir Starmer and his team are planning the most immoral and evil government in our country's history. By refusing to engage in an emergency decarbonisation of the UK economy they will play their part in locking in the destruction of our civilisation over the coming two generations.

Imagine a massive block of concrete at the top of a high building - it is about to be dropped on an inescapable cage filled with our children. Labour's policy is to allow the people with the block of concrete (the economic elites) to drop it and shatter the bones of those kids. Imagine it. Not just this on one day, not just the following day, but every fucking day forever and ever - like some deathly Greek myth. This is Labour policy: the infliction of endless torture and pain. That is what you are voting for when you vote for the Labour Party - mass starvation, mass slaughter, mass rape. A Hothouse Holocaust. Forever.

This is what it means to allow the world to go over 2C - to create a billion refugees, to allow the arctic ice to melt - to destroy our ability to grow food, to allow the Atlantic current (AMOC) to collapse - to turn this country in a barren desert of ice, to allow sea level to rise - to destroy every coastal city, to create unbearable heat for everyone in the subtropics - to kill them in six hours, to melt every glacier and all the snow in the mountains - to deny billions access to fresh water. Then there is the Amazon, the methane, the ozone, the acidification...the list goes on and on and on and on and on.

Labour plans to have all this happen and for it to happen all at once. Don't vote for them. Don't repeat the terrible error of voting for the Nazis in 1933. 12 years later 10 million Germans were fleeing the Soviets, 2 million raped women, and the country was utterly devastated. Labour's plan is for this to happen to you and your children. Except this time it goes on forever and ever and ever.

Don't delude yourself by the fairytale that decent upright people cannot create such evil. Decent and upright people stood by while millions burned in the concentration camps. Decent and upright people ran the European colonial empires. Decent upright people owned and killed their slaves. Decent and upright people create Hell. Then and Now.

Instead, follow the facts - follow the reality -the science, and don't blame the messenger. All I have said here is confirmed by 10,000 peer-reviewed papers. Emergency action is needed and Labour will not enact emergency action. These are the only two facts you need to know. Being upright and decent is irrelevant. Starmer will stand by. He will do nothing.

As it happens when the election was announced yesterday I was watching a clip from the greatest horror film of all time - Threads. Watch it on Youtube - watch the nine-minute clip: the bodies, the rubble, the blood, the faces, the shit, the tears, the end... of everything.

This is what 2C means - what do you think 1 billion refugees (for starters) looks like? Get fucking real. Don't insult yourself.

Then look in the mirror and say this:

> "No way am I voting Labour".

> "I will stop being a slave to our executioners."

Independents are standing to create system change - our only hope - vote for them. If you don't have that choice write on your ballot paper - "they do not represent us". Then go home and prepare to do your duty to resist the greatest shitshow... ever.

If you read to this point, I wish you all the best. It's the least I can do.

Image by Oskar Pierre Castro

Political Violence and Responsibility

In the wake of increasing violence across Europe on the political class, we must remember their own violence against the people.

(May 16, 2024)

Democracy has two duties - to protect the people from the rich and to protect the people from death. On both counts the current political class is failing. Whilst they protest against nihilistic attacks on their own members, they are also responsible for the most violent dereliction of their sacred duty - the duty to protect.

For a generation now they have allowed the people to be subject to the violence of the rich - the remorseless decline of public services, the increasing mechanisms of social control, the nauseating moralism that if we work hard and keep our head down we will prosper. The violence of this endless bullshit.

For a generation now they have allowed the people to be subject to the violence of death. The ultimate disgrace is our politicians' willingness to trade the end of our societies for the private interests of the carbon lobby - the fossil fuel interest, the agricultural industry, the savage endgame plunder of our world by the forces of capital.

People can only take so much of this reality. Most buckle down helpless and hopeless, some resist and are thrown into prison, others take refuge in conspiracies, and a few turn to violence against themselves. Others direct it against their tormentors.

The irrationality of this obscene rise of political violence is a mirror image of the irrationality of our obscene politicians. They are an outrageously infantile and cowardly class that continues to administrate the destruction of all we love. All they offer is the final insult - a list of pledges we all know will be broken. Like those before, and those before that.

What do they take us for - idiots? Sheep to be led to the slaughter?

As Martin Luther King said, we have a choice: Nonviolence or Nonexistence. Some are us are fighting every day for nonviolence - with the power of Powerlessness and the power of Love. But if those in power continue to facilitate this violence upon the people - then they will increasingly come to experience the meaning of the old saying:

"If you spit in the wind it comes back and hits you in the face".

COP 28 — Run by Fascists. Legitimised by Liberals

(December 6, 2023)

Let me make a well considered statement:

The COP process is dedicated to facilitating the greatest holocaust in human history.

Any organisation that participates in this process in 2023 is guilty of complicity in that holocaust - the world on fire for ever more.

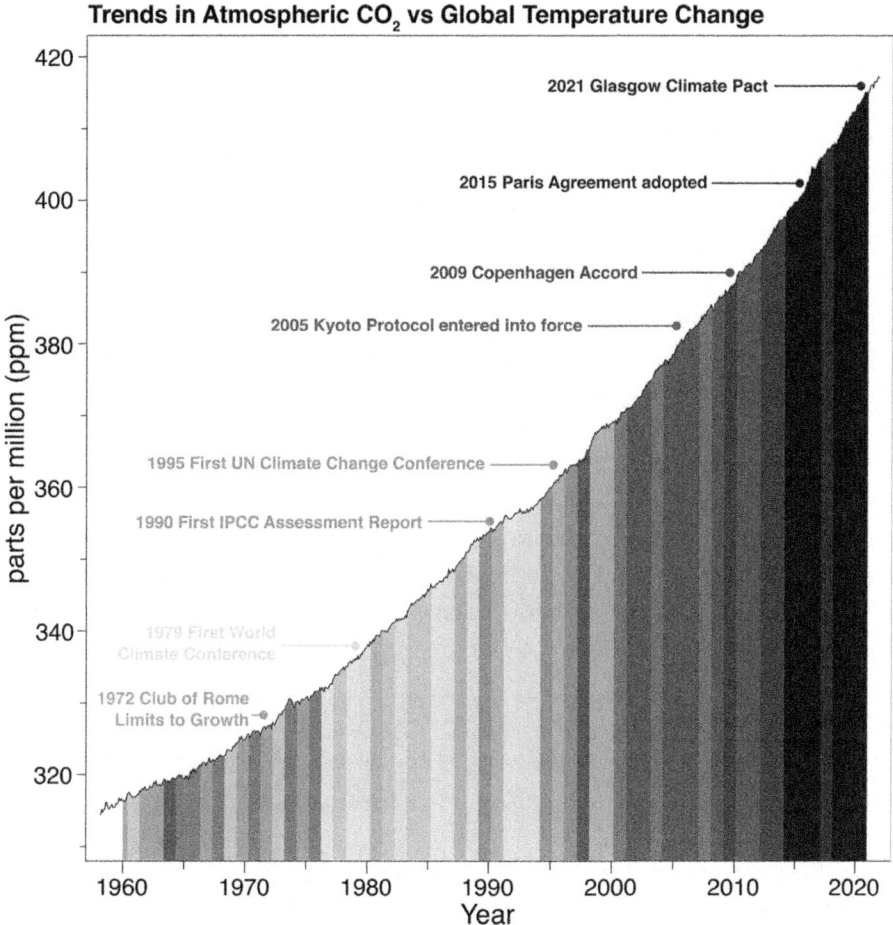

Trends in Atmospheric CO$_2$ vs Global Temperature Change

Look where 27 COPS have got us.

We're at record high emissions and the current COP president says there's 'no science' behind phasing-out fossil fuels. It's a joke. A mockery of humanity.

If it's a mockery then why continue to go to COP? You just make a mockery of yourself.

Would you have gone to a peace conference with Hitler in 1939?

It's a 2020s version Soviet show trials - anyone who engages with it is a "useful idiot" for the greatest genocide project in world history. Even

Greenpeace has voiced qualified support for COP as "not only a historical opportunity but also a stage to demonstrate the UAE's diplomatic power"

Meanwhile so called "campaigners" are criticising Rishi Sunak, the King and David Cameron for taking separate private jets to get there. If they took the train to the mass murder conference would that be okay?

Winning "winnable" demands at this stage is just another variation of the global suicide project. Only with a strategy of resistance to change everything will we have any chance to save anything.

IT'S TOTAL FUCKING BULLSHIT.

How can anyone with a semblance of realism or morality financially support an organisation that gives credibility to a conference that facilitates the greatest holocaust in world history - or what the fuck do they think 1 billion refugees means?

It's time to drop the corporate frame "climate", the murder weapon, and instead focus on the murderer - the elites. Either we collapse into fascism and extinction or we create a progressive revolution.

To be clear - three things are needed if we are to avoid collapse into effective extinction.

1. The total decarbonisation of societies
2. Massive investment and roll out of earth repairing/geo-engineering technologies
3. A wholesale political revolution which puts the common interest before corporate interest

The first two will not happen without the third.

The only way to force social changes at scale in the small timescale we have is through Civil Resistance.

Artwork: Bacon, Francis; Figure Study II; Kirklees Museums and Galleries

Just before the Beginning of the War

(Sep 24, 2023)

Recently British Judge Silas pulled posters off the walls outside his court saying "Silas Silences". He had been imprisoning people for saying "climate change" in his court. The police had not followed his orders to take them down apparently. He no doubt was finding himself in new territory. We all like to stay safe but then suddenly the cold wind drives in and we're swept up into a new world – bewildered we stagger around trying to grasp a breathe.

Recently I suggested we hold placards outside his house – I pictured it – dreamy green detached, along a home counties road, then down an enticing lane, not too long but long enough. Maybe we could sit in the drive and he couldn't get out. Maybe we would go to his daughter's school and hand out leaflets saying – someone's dad at this school is jailing people who are trying to stop the mass murder of millions of people like you. Innocent like you. Would that be cruel? But necessary. Cruel to be kind.

A question we still have time to ask before the war closes in.

In prison I thought I might like to write something beautiful – something gloriously without any "point"– as an act of defiance. There is something strangely satisfying about a random act of beauty – like kindness in a war zone. I would have reach deep inside myself to find such beauty and maybe nothing would come out. Sometimes that's the way things are – you come to learn that. But I like to think – there could be a little light in the darkness before there is only darkness.

Or maybe I am being too pessimistic. An infinity of parallel universes stretches out before us. And in at least some Mr Silas is right – it turns out we were just a temporary bunch of neurotic idealists after all. It all turned out okay – capitalism conjured up its magic one more time. Necessity is the mother of invention old boy. Nothing to worry about. Like the Germans in 1945 – Mr Hitler will come up with the techno fix – a new missile that will protect the Homeland. So they could stay free from the burden of having to think a little longer.

In some public schools during world war one only a quarter of the boys came back from the front in one piece – alive. It sometimes turns out there was plenty to worry about – and then much more.

Many parallel universes must end in the darkness. Unfortunately it seems we most likely will be right.

In the meantime George Monbiot has said the police "can do anything to us". Well, it depends what "us" is. But what a question that is. No one is allowed to write about philosophy in the guardian. There's no "likes" in such introspection. We English don't want to look too deep. At least not yet.

Nietsche apparently never sold more than 300 copies of any of his books during lifetime. Poor thing. It was only after the first world war (the one that broke millions of boys into pieces), when liberalism seemed like an eternal very bad joke – that his books took off. When people started picking at the scab, and found solace in misery.

But we are not there yet. Again no one reads Nietsche and why would anyone want to create anything pointlessly beautiful. We have so much – still, why do we need some superfluous beauty as well.

Till everything turns. And then. The heavens will shake with such anguish. Heaven knows they will.

You get up in the morning and do what you can. The decision to love is always a hard path but I know it is the only one with any real meaning. As long as it remains a pure idea. This world is too harsh for such a thing to come up from within it. There are two worlds, the cities of man and god, as traditional theology teaches. Not that anyone reads theology yet either.

And if that means I have to suggest we leaflet the children of judges – well no doubt any moral scruples will soon be forgotten in the deluge. When these children ten years from now are sent off to shoot down black bodies crossing the Mediterranean. And pick up pieces of bodies in the poorer parts of town. When the war begins. The one that will never end. Getting a little upset by a leaflet will seem an infinitesimally small thing.

So best to try and write something beautiful then – or at least try. A comfort before I go off to work.

To a world where no one wants to read half decent prose about the end of the world. Loser.

Like Nietzsche. Get a job.

Spiritual

Dark nights of the soul. Ecstasies of being.

Crucifixion: William Blake (1805)

Telling The Truth: The Easter Spirit

(Mar 31, 2024)

I was asked this week to step down from a group for saying unpleasant things on Twitter. At least I think that was the reason - the English way is never to be clear about reasons. I could (and briefly did) get into a bit of rage about it, but if there is one thing that the last few years have taught me is that people really cannot handle the truth - perhaps for good reason. Live for today - what is coming down the line is deliberately made hazy. Messengers who shine a light have to be killed - at least in an English way.

But what enables me to get through the day in this mad, mad world is the idea that I can write something real - that which appears real I might be prompted to say because in this culture nothing is allowed to be actually real. It's all a matter of opinion for those who have the privilege to think that way. The train coming down the tracks is just one of many things. The joy of endless choice. The madness of endless freedom. The world is what you make it.

I find it slightly hilarious how at this moment so few people want to write about what is actually going on. I think there are less than five people right now who can be emotional, use ordinary language and take things even halfway to their obvious conclusion. Obviously we are heading towards a holocaust so much worse than the previous one - and no, not in a million years am I trivialising - nothing could be further from the truth.

*"**Don't mention the war**"* my friend who asked me to step down said a while ago - the essence of English repression. *"**Don't talk about Eichmann**"*, another old punk friend tells me - something has been lost from his soul. And definitely don't mention hanging.

It is indeed hilarious, and so so sad of course, because I say such things to prevent suffering - to alert people to the horrors they are setting in play for themselves and others. I would never want to hang anyone - I'm a follower of Gandhi for fucks sake.

Image: Damien Hirst, False Icon (Golden Calf), taxidermy in 18 carat gold and bulletproof glass vitrine, 2008.

Finding Our Religion

(Aug 22, 2023)

Preamble: On God

I have been speaking a lot to the theologian Carmody Grey. She's asking what I consider to be the most explosive question of our time:

> "Why is it that we are the most advanced, knowledgeable and well-educated population in the history of humanity and yet we have utterly failed to stop the locking in of the greatest episode of human suffering and injustice in history – catastrophic climatic and ecological breakdown?"

Camody's Talk

If you are tempted to give shallow answers to this question I would encourage you to ask yourself the question several times until your answers start to make you deeply uncomfortable. Because the real answers to this question are nothing less than a devastating critique of not only our way of life but the very way we look upon our "selves" and the "world".

I chose the title of this post in opposition to REM's song "Losing my religion". These three words sum up everything that is going wrong. We have assumed that religion is a personal matter ("my" religion) – it is not. And we assume it is possible to "lose it"- it is not.

The reason why we have failed to act, Carmody suggests, is that our dominant secular rationalist paradigm is obsessed with the nonsense that people change due to being given information: just giving people more facts like a computer – 30 years of them. In fact (as it were) people only engage and act when they realise their core values are being violated. Facts in themselves mean nothing. They are just numbers. What we need to start talking about is the Sacred.

She suggested another question to me last time we met. "To what do we owe our deepest loyalties?" – a seemingly straightforward enough question but it is packed full of existential dynamite if you find the courage to keep asking yourself this question. In the early days of XR, we had a go-round in a press spokesperson training session. People had to role play their response to the question (no surprise) "why are you disrupting the public". We had several go rounds. At first, people gave superficial generalised answers. But within ten minutes people were in tears. They got to the bedrock: "it's because I love my younger sister so much and I can't bear to see her suffer".

To what do you owe your deepest loyalty in a time of social breakdown? You too will be in tears once you answer this question truthfully.

The point is this. You cannot rationally or empirically justify your deepest loyalty – loyalties emerge, you sense them, and you are them. They come through you. They are - admit it – non-rational and yet they are all we have.

Having them is what keeps us sane. Without them, we fall into the abyss of the destruction of ourselves/the world.

Some people suggest we need to dabble in a bit of "religion" the way bureaucrats want to dabble in a bit of "sustainability". The days of dabbling are long gone. The Freedom Riders did not dabble. They did not stop in Virginia and get a little ticking off. No, they went into the heart of darkness – the most fascist state in the US, Alabama. They were beaten up, hospitalised, and imprisoned. Everyone hated them. They exploded into the obscenity of racism. That's how you change history.

To save ourselves, to change history in 2022, we have to explode into the obscenity of materialist individualism – nothing less. People do not even see what this is – just as the white population of the American South did not "see" racism in 1961. You do not see something you swim in and have always swum in.

To get people to see their materialist individualism (this world is the only world, and I am real and the only important thing) you have to declare you are going to do nothing less than to create a new Religion. You have to publicly declare for God, and that those that go against the will of God will go to hell. You have to drive into the dark psychic heart of the putrid "rationalist" hubris that has us standing by while the world burns.

For 200 years or more we have all been subject to the diabolical frame behind the question "does God exist?" To decide to answer this question is by definition to accept a Godless world – a world without enchantment, mystery or awe. This is because the very presumption of this question removes the possibility of anything other than either "existence" or "non-existence". This rigid binary is the most destructive nonsense in our culture today.

"Does our "self" exist? Yes? So show me where it is – give me the material evidence. Show me the damn thing. You stupid idiot – how can you believe in the self when is it nowhere to be seen? It cannot exist."

"Does the past, the future exist? Yes? Where is the past – show it I damn you. It is nowhere. It does not exist."

As soon as you separate the world into "existence" and "non-existence" you are forced to destroy everything of value, everything that gives meaning - the imagination, the emotions, the sensibilities, the sense of looking at ourselves – you destroy life itself. Where the fuck is "life"? No one can "show" you "life". According to materialism, "in fact", there is nothing but dead matter - the brutal world of lifelessness. This is why the most secularised "educated" populations in the world sit on their arses and don't give a fuck – because they can't even see any life to be saved. They can only see stuff. And stuff is dead.

God "just is" – so get off my back. Stop stressing about it. Stop asking if God exists or not – stop being such a fundamentalist literalist. Like the word "self" – it is a name for something, something which cannot be named. We "know" many "things" that cannot be named, that cannot be seen – existence or non-existence does not even come into it. We expelled God to pretend we can do away with a pluralistic sense of self. We made the self an atom and the world a world of atoms – hard, separate, dead. We are destroying the world because we are already dead. We killed our "selves" when we killed "God".To go into the heart of darkness in 2022 is to kill the killer of God. Only once we rescue God from death can our souls breathe again – only then can the story of life re-begin. Can enchantment return? Can we return from "facts" to stories? There can be no life, no soul no conscience without God – meaning without a sense of the divine, an awareness of the sacred. All these sensibilities are a team and each element is essential for all the rest.

Once we rescue God, once God returns to the story, then can we again become whole – whole by submitting to God. God is us and we are God. This is not a fact, it's not a debate, – it is a story, and it is no less "real" for "just" being a story. Open yourself up to this world of many dreams before you die before you have even started to live.

It's the same with Religion. You can't have a bit of religion – it's either/or. It's faith or eternal darkness. Without faith, there can be no reason, no sanity, no compassion. Faith is the decision to believe, knowing that there can never be any ultimate rational foundation for either metaphysics or ethics. We do not maintain our decency by logic but by intuition. Balance is a function of collective sense-making not a series of deductions that always ends up sending us into the abyss of nihilism.

The question Carmody asks is not "do you have a religion" but rather "what religion do you have". Not "do you love" but "what do you love". And on the answer to that question, God will tell you who you are. You can't serve yourself, that's a contradiction in terms, you can only serve Good or Evil. On some things, you have no choice but to choose. And this choice is the most important choice of your life.

Once you emerge out of the fishbowl into this new way of seeing the world – armed with God, armed with a Religion of Life – the sense of freedom is immense. The world is spirit and spirit by definition is free. Materiality was just another myth like any other myth. We are here in this life to master ourselves and the world is no more (or less) than a stage on which to enact this mastery.

My chat with Carmody

The advantages of God - The Dogma. The love.

A friend of mine looked at my plans for a new religion and said it was "fun" but he did not like the "dogmas". He thought it was fun because he thinks you can have religion without dogmas. Dogmas are as essential to social life as air is to biological life. Dogma is that you don't rape your sister. Dogma is that you don't kill your babies. The oldest dogma is that you do not do unto others what you would not have them do to you. In modern society this dogma no longer holds. The modern dogma is you do nothing while creation burns.

Anthropologically speaking dogma is a means by which a society can maintain itself. It can allow no argument because argument opens up the possibility of contention and contention can only lead to dissolution and thus an existential threat to society.

Societies go through convulsions as a dogma which is no longer functional has to die and a new dogma that serves the new objective conditions emerges. One regime, in the widest sense of the word, has to die for another to come into existence. A society often literally has to die for a new one to emerge or take its place. Now, for the first time, there is only one society in the whole world. It has to die to give birth to another society. Or it will literally die in the flames of Nature's revenge. In its murderous hatred for itself.

A new set of dogmas – a reinvention is necessary. The advantage of God is that it allows for transcendence – the life-transforming realisation that we are not our "selves". Without this, we are doomed before we start.

Dogmas are metaphysical, not ethical. You do not do unto others as you would have them do unto you because it is right, but because to do otherwise is the ultimate transgression of what is it to be who you are because to do otherwise is a violation of the sacred.

You do not do the good to be good. You do the good because to do good is to enact the essence of your consciousness. You are good, creation is good, and the world is good. This is a dogma. It cannot be denied otherwise the whole edifice of humanity collapses. Some things cannot be questioned.

Resilience is rooted in this self-knowledge, not anything as impermanent as "ethics".

In these End Times, the world can no longer be the centre of the world – people can only take much of "reality", meaning the material world, as was TS Elliot's assessment.

The world has to become the world of the spirit – idealism replaces materialism, "God" replaces "Man".

Resilience is rooted in the collective, not in the candle in the wind which is the old fragile consumerist self – ever fearful because everything outside itself is separate, disconnected, and thus a threat. This is the hell of narcissism.

The central dogma of the new religion, the new regime – is this:

> Only in service to the Good can we become what we already are.
>
> Only in Love – defined as acting to promote the well-being of the other – do we express the essence of what it is to be human.

We are not a blank slate. We are children of God. A half beast half God. In eternal suspension. The nature of this reality was the central focus of cultures for thousands of years. Only the arrogance of our sliver of time, called modernity, has led us to believe it is no longer important.

This is the reality: We are not alone. We are intricately bound together with others, and in that togetherness, we enter into communion with God. The modern lonely "self" was invented. It can be un-invented. We can return to our common purpose of attending to the well-being of others.

In other words, the deepest and eternal dogma is this: We Are All One. This is the final realisation. We were all one all along. And always will be.

Love is incompatible with standing by in the face of injustice because injustice is a violation of Love. There can be no separation here between the private and the public. Love – the essence of Love – is an act, the act to combat Evil which is the opposite of Love. Love is connection. Evil is disconnection. Love is the process of creating a connection that comes through the disruption of Evil, through the struggle with Evil. True love is manifested as disruption and struggle – it is militant and uncompromising. All forms of quietism and fatalism violate humanity because they attempt to abandon humanity. Humanity is One – it is a whole – you cannot insulate your "self" off from the rest of humanity, from the evil that is happening around you. Love is direct action. Until there is only Love there is no Rest.

The reason for Evil is for God to answer not us. The calculation on whether we win or is not for God to answer, not us. Our job is only to become who we are – that is, to step into the path of God and remain on that path. There is no separation between who we truly are and Love – and there is no separation between Love and the overcoming of Evil. Love cannot step away from looking Evil in the face– it can never live in the false comfort of temporary privilege.

Love is manifested only in the act of overcoming Evil. This is the practical program of the new dogma.

We can give all other questions to God. We are not here to reason why, we are here to love. We do not need to know why things are the way they are, or whether or not we will win. That is the path of humanism, or separation, of burnout, delusion, or despair. All these things can be given to God. God in other words sets us free.

Without God, when we look at ourselves in the mirror we see, to our horror, only ourselves. We must take on the full burden of this world. Very soon we will not be able to bear this burden any longer – our souls are not designed for such a weight. Our souls are designed to rest in the lap of the Mother. That is why without God we will be lost in suffering as this world burns. Whereas in devotion to God we return home to the way things should be, and to the way the world truly is. Our role is to Love and let go of all the rest. If our love brings about the Revolution so be it, if not so be it. Either way, the stars will still be shining in the nighttime sky. God is always with us whether we acknowledge it or not.

This then is what I mean by finding our religion. Giving up on the materialist delusion – already shattered by modern physics, echoing the lost mystical traditions. Giving up on the masochism of narcissism in a time of social collapse. Giving up on intellectualism at a time when the hubris of our plan to go it alone without God, has clearly failed those with eyes to see. God is waiting, and with God by our side, we will take down the Empire of Evil. Or die trying. This is the real meaning of "the fight for our lives".

Transcendence and Disruption. No Regrets.

To Act is our Religion.

People say that you cannot get yourself to believe something even if you think it would be good to believe it. I used to believe this but I recently met a rich guy (very rich). He had got divorced and it had driven him into a spiritual crisis and he made a decision to give up on the material view of the world. Everything instead he decided is consciousness. We spoke about the three D's – divorce, disease and death. All of them make a mockery of the idea of a life of meaningless work and consumption. Now he is going to give his money away to climate action.

I'm with Pascal – take yourself into the ritual of religion and you soon find yourself absorbed into the enchantment of the world of spirit. When you face death you pray to God, whether or not God "exists" becomes an irrelevance.

As the material world becomes ever more hellish due to climate collapse and its social effects, belief in it – the religion of materialism - will become ever more unsustainable. Some will turn to hedonism, others to sadism, and others to escapist quietism. These religions will not save us. To find our true religion – a religion which will save the world - we need something like what I laid out in the previous two posts. The entry point is the rediscovery of God – as part of a family of similar rediscovered sensibilities – the good, conscience, consciousness, awe, enchantment, and spirit.

And we find these sensibilities in the only place they can be found at present - in the act of social revolution.

This religion is based on acting for the Good: the pro-social intolerance of evil – the obscenity of injustice. This can only be done when we release ourselves from the ego: to find meaning through public disruption of the empire of death – and give up the old religion of stuff and deathliness.

It is proactive. It aims to take over the world.

What is laid in these posts is not a thought experiment – a fringe project. It is THE PROJECT. It requires organisation, hierarchy, money, and strategy. It is a collective material project which comes out of the rejection of the material. This paradox is at the heart of the genius of its transformative power. Instead of being a religion of the material which destroys the material (i.e., the present system), it is a religion of the spirit which saves the material. Only when you lose yourself (enter the spirit) will you find yourself (save the world). This religion will move into politics and take over the state. It will take over economics and create a steady state economy. It will take over social relations and create a culture of universal rights. It will be the salvation of this world by transcending this world. Only when you embrace this contradiction will you be able to overcome the pathological contradictions of our politics, economy and society.

The details will have to be worked out. What is written here is a draft. But the details will not be created in the academy, not the quiet retreat, but on the streets, in the sites of social struggle. There will be no monolith but neither will there be a chaos of initiatives. Success resides in a middle way – a close ecology of social formations in which half cooperate and half compete to find and then roll out the new social and psychological technologies of the post-carbon age.

This is no escape from the world – but at the same time, we are not of this world. The world's a stage and as such, we need not be attached to it, and this is how it will come to be saved. Humanity is to be invited into an adventure of the spirit. This is its Destiny, in partnership with that part of itself that is God. The Divine. The Pure Spirit.

I am working with others on launching this religion – its structures, its rituals, its transgressive agenda. It may not take off or it may create the grounds for another initiative that does, or it may go nowhere. No matter. I am not attached. We are here to enact the spirit within the material. We are already free and in this freedom, we have the best chance of releasing ourselves and the world from eternal death.

This is what it means to be truly alive at the present time.

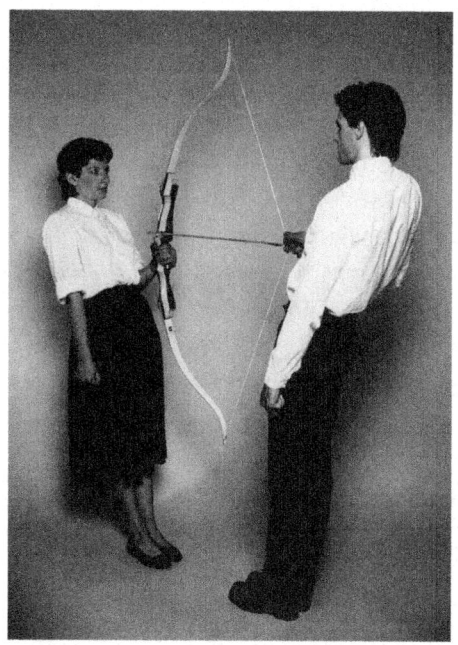

Photo: Marina Abramović and ULAY, Rest Energy, 1980.

Spirit = Action

(April 10, 2022)

One of the most obscene ideas is that you can be spiritually special, progressing on the road to some enlightenment, while standing by and allowing the world round you to get utterly fucked. Yoga outside Auschwitz will not cut it. No, it just won't.

I work hard to find much compassion for people who promote this idea. Maybe some time 20 or 30 years ago, they could have got away with such gracelessness. When the world was just average, ambling along bad. When capitalism was on an up day. When the world was getting high on the novelty of infinite connectivity. Who would I be to criticise a little harmless neurotic indulgence?

But now? No.

When I was teenager I sat on the pew of my Stockport Methodist church each Sunday, rage surging through my veins at the weekly betrayal of Jesus. The drug of my passion surged through my mind carrying me off – I was helpless to stop it. Jesus said we should be poor and then I looked at the congregation. The ultimatum appropriation – the church's claim to over Christianity. I was off my head, spinning, dizzy, with all the newness, my body, my mind, this world – this world of suffering, of injustice, stabbing me to death each day. Rest was a foreign land I had never heard of.

I have always felt a hatred of conventional "spirituality" for that reason – my hatred is deeply flawed of course, as hate always is, not least because the feeling itself is excruciatingly spiritual.

I took 20 years off to cool off. Carrots in the field. Nappies to wash. Stuff.

But, returning to the fray, I soon started to wobble, despite two decades of practised repression.

I tried to do my best within XR. "Regenerative culture" "self-care". Okay whatever. But what did it actually mean in practice? Feel bad, take a rest. Love yourself. Ban Nietzsche. The outcome: distraction, disintegration, depression - burnout.

Is this harmless? No. Because we are here now, not 30 years ago. The good times have gone, the slack, the self discovery, the me me me. Now we all face hell.

I try to be kind. Live and let live. I am no longer young, not so extreme. Take the middle ground Roger, I tell myself. But what does this mean when 500 million Africans are lying in the dust in a decade's time. We are firing that mass murder weapon today to manifest this outrageous obscenity. In a decade the devil's work will be done.

The devil is us.

What does moderation mean in a time of mass killing? To walk on by? This will be the eternal question asked of the 2020s. The decade when we finally threw it all away.

The paradox – there's always the paradox – is that the spirit does not live in "spiritual places", the quiet retreat, the calm mountain view, but in the site of confrontational action. Any "spirituality" that undermines such action is plain evil. Regenerative culture, if it is to mean anything real, is only found in the agony of courage carried through into collective transgressive civil resistance. The body over the mind. The spirit over the body. The community over the individual. Drama over "contemplation".

"That was the best two weeks of my life" as I like to tell people in my evening talks – that's what was told to me many times by people referring to their outing from complicity into truth – otherwise known as arrest. The journey, in community, from cowardice to courage, from self obsession to service, from indecision into action. From a "safe" place to the real place.

This is how it seems to me - sitting in a prison cell is a spiritual act, the only place of rest – not sitting in the pew, not in the yoga class, but when looking at the intricate roughness of the chipped wall day in day out. Here is God. In a Time of Evil we know deep down we will only find him here. So silly to have pretended otherwise.

Civil Resistance

It's our best shot at survival.

".. then they fight you, then you win." – Gandhi

Why Repression is the Best Thing Since Sliced Bread (Analytically Speaking)

(Sep 24, 2023)

Repression – arbitrary arrest, prison without trial, violence, lies, manipulation -the whole paraphernalia of the pathological elite is disgusting and outrageous. And I can say this from some personal experience, having been banged up for months for making a speech. But as battle-hardened readers of my posts you know I am not going to write on something so intellectually lazy as why the bad guys are bad guys.

No, what we need to do is to look with laser focus at the rapid descent into repression of the UK carbon regime to deeply understand why the present moment – this getting towards our darkest hour – is our moment of greatest opportunity, particularly if we play our cards right. Let me say this: this moment gives us the greatest probability of overwhelming the UK government than at any time since the Rebellion of April 2019.

How come? It comes from the most exciting word in the English language: "Backfiring". Many of us have a vague understanding of the phenomenon but it is fascinating to see the dynamics play out before our very eyes. Like being arrested for wearing a Just Stop Oil T-shirt. Did you see that right? Yep, people are being arrested for wearing T-shirts – not as a result of some mix-up – some Met fuck up. But as in the bright light of day, an act of police policy. Nothing "went wrong". Add that to the getting imprisoned for making a speech (me), getting imprisoned for saying "climate change" in court, and getting sent to the Old Bailey for holding a placard saying juries have the constitutional right to go against the judge in their verdict, added to the trail of police raids and arrests for ...er... being in the wrong place at the wrong time. Suddenly, what has happened to the black and minority communities for decades is going mainstream.

So first let's look at the basic dynamic: when the opponent uses repression towards nonviolent resistors they create an instinctive sympathy from supporters, observers and even opponents of the cause. This can be conscious, subconscious (grudging respect) or arguably unconscious. In principle the greater the aggression, violence and/or punishment – the greater the sacrifice – the greater the moral attraction. This is backfiring. This for instance works in a simple case of seeing others you know getting arrested – then in a group discussion later you are far more likely to sign up to engage in civil disobedience to the point of arrest yourself. Action is worth a thousand words – it's so true. Seeing is believing. So the harder they push the more the backfiring potential. This is not deterministic – it doesn't happen every time with everyone – but the odds are high.

But there is something else going on which explains why semi-authoritarian regimes like the UK are most vulnerable to the civil resistance methodology – mass civil disobedience enacted day after day. In liberal regimes, you have to do "quite a lot" to be arrested if you take part in open civil disobedience

– and even more to be sent to prison. Sitting down on the road as they did recently in liberal Holland for instance – led to arrest but no charging. To end up in prison requires doing something dramatic – "criminal damage" to a building or an oil refinery – an "activist" type thing – certainly not what ordinary people would do and not where they would go. When people do "activist" actions they create a little emotional simulation in the public because people think "Oh it's those people doing strange activist things – that's nothing to do with me." They switch off. That's the main reason why the press does not cover doing bad things to bad people.

In contrast, in semi-authoritarian regimes the "site of confrontation" as you might call it becomes the "everyday action" in the "everyday place" – most especially marching down a city road. Doing this moves from a non-arrestable activity to something to be pulled off the road for, handcuffed and dragged off to a police van. The optics are different – this violates the "natural" sense of justice of ordinary people – why is that happening to people for just marching?

Similarly with the T-shirt incident – a secret action on an oil refinery gets no publicity and little sympathy – while people get arrested for standing around wearing a JSO T-shirt – what the fuck. It has become a national talking point.

So "repression" in actuality, means making the everyday illegal and so it vastly increases the backfiring potential.

Lastly the "entry cost" is significantly lower cognitively and materially – you can see yourself doing a march – in a way that you cannot see yourself climbing over fences, and it's easy – you just go and join in down the road. So lot's more people get outraged and a lot of them decide to join in because it's easy. It's a march and you get the numbers. It becomes a people power thing.

This is how classical civil resistance works – it's not complicated – you go on the street and stay there till the regime reacts or falls. It creates the opposite of what was intended. Handy!

Of course, the opposite dynamic applies, repression can create disengagement, as the pessimists like to remind us in their self-serving fatalistic analysis – so there is nothing that can be done. The Police Act "prevents protest" as if we humans are bits of a car engine – if you don't put petrol in it the car won't move. Humans are not engines – we are souls with a moral sense. We are violated by the violence of power. And if we as leaders and designers act in smart ways in service to our communities we can bring about the glory of collective action. It's happened before it's going to happen again. Sooner than we all think.

Just watch what's written on your T-shirt!

Outside the London Mayor HQ

Life's Trials are all Heaven's Pearls

(Jan 16, 2018)

About this time last year I did a post which said I felt like I had to start doing direct actions again (rather than just researching them).

Since then I have been arrested 8 times (7 nights in police station cells), been suspended from King's College twice, done two hunger strikes, spent a week in prison and been fined £1000 at various court appearances.

Well someone said I shouldn't put this up as a post as I would be showing off but really I don't think it's a big deal – millions of people around the world last year suffered many times greater hardships in the pursuit of justice and sanity than I have. The point I wish to make is that last year was probably the most fulfilling year of my life so far, not despite all this civil disobedience but because of it.

As I finish my PhD research into creating effective radical activism/political change I am increasing convinced that only "high end" disruption and sacrifice can be effective against the forces we are now up against. And so I am increasingly interested in the cultures and life philosophies which create the ability to undertake these actions. I suspect that this means looking at traditions and modes of thought which are usually mocked and ignored by secular western progressives. For instance I think Levon Helm's lyrics point to a wisdom which personally inspires me.

Life's a strain of painful things to be overcome
We mark the mile by our trials and suffer everyone
But hold them near and keep them dear and don't be ashamed

Trials of the world are all heaven's pearls
Trials of the world are all heaven's pearls

We gather scars from tangled bars that catch under the skin
The friction burns as they turn working deeper in
But day by day sharp edges fade, smooth, burnished and fine

Trials of the world are all heaven's pearls
Trials of the world are all heaven's pearls

I'm planning on this coming year bringing along a good few more pearls

Prison

My writings from the cell.

They Came For Us and Soon They Will Come For You

A Letter to the British Left on why we must unite against the encroaching fascism of collapse.

(Aug 20, 2024)

> **Note:** *This letter was written from Wayland prison and originally published in The Morning Star.*

I was wrong. When the recent trial started for what is now called the Whole Truth Five (#WTF), I naively thought I would walk free, that I would not be found guilty. My crime was a 20-minute speech on why people should join a Just Stop Oil non-violent Civil Disobedience campaign. Or if I was found guilty, I reasoned the jail term would be no more than two years. I had already spent four months in prison on remand followed by a year and a half on a curfew tag so a two-year custodial sentence would be no big deal.

Three weeks later, after I had been dragged out of the Court and into the cells several times for speaking to the jury, about evidence the judge deemed "irrelevant" I was sentenced to five years in prison. Repression is not a gradual process, it leaps out at you and takes you off guard. Do you remember the Solidarity leaders in Poland? They were invited into talks with the Polish government but when they got to the meeting, they were arrested in one fell swoop and imprisoned for years. You don't think it will happen to you and then it does.

Just Stop Oil has had an uncomfortable relationship with the Left. There is a pretence that it can be ignored, but Just Stop Oil reminds us that the future will not be like the past. In a decade or so societies will be falling apart as we pass 2C of heating and experience extreme weather events which will push hundreds of millions of people into destitution and migration. It is going to be an Almighty shitshow, and it will not stop. You can't negotiate with physics, with a thousand peer-reviewed articles. Just Stop Oil reminds us what resistance, that far-off folk memory relegated to Netflix, actually looks like in the present moment. Thousands of arrests, hundreds of imprisonments and a 5-year sentence for making a speech.

It foreshadows what will happen to the left, soon enough. As Trotsky says "You may not be interested in war but war is interested in you". There is not going to be a bright sunny Green New Deal, the neoliberal fantasy of the Labour Party. There's going to be a rupture, and then a rapid descent into poverty and starvation. Along the way the ruling class will have the police knock on your door and have you put away for half a decade. Capitalism is not just destroying our prosperity, it is destroying our very means of existence. To take just one example, the Atlantic Meridional Overturning Circulation (AMOC). This is the earth system which warms the British Isles, delivering warm water from the equator to gently warm our shores. It will collapse at some point before 2050, most likely around the mid-2030s. This will reduce temperatures in the UK by between 10 and 30°C and end arable agriculture.

What will that mean for British workers? For British farmers? For Europe, for the rains in Africa and Asia? Just Stop Oil is not a sideshow it's raising the alarm about the biggest fucking disaster in human history.

What needs to be done? It's time to unite the non-Labour left and the climate resistance space, to fuse. We have a lot to learn from each other and millions of people are crying out for initiative and leadership. A program for decency, compassion and survival. A massive reduction in inequality to fund the reduction in emissions and earth system repair. Anything less and we will have nothing but fascism and mass death.

I wish to thank the Trade Union leaders who along with 1,200 others signed a public letter to protest the sentences of the #WTF. This must be a beginning. We need a series of joint meetings, a national tour and a new politics using assemblies and dual power.

My Trial Was A Sham

This is my closing speech to jurors in the Whole Truth Five trial. Afterwards I was found guilty and given a 5-year prison sentence.

(Aug 16, 2024)

Note: *This speech was transcribed from a prison phoneline. You can listen to the audio version on Spotify/Apple Podcasts or watch the Youtube video with subtitles.*

I would like to start by acknowledging that this has been a strange and difficult trial. As the judge has said, he has never had defendants dragged from the court and put in prison, and it's not usual for the jury to be sent out so often, prevented from hearing what a defendant has to say. I want to apologise for the commotion; it is not how I would like a court case to proceed, but I'd like to draw your attention to a very important outcome of this disruption, namely that the judge was persuaded to change his mind and allow you to see four relevant, agreed facts about this case—about the harm and the criminality that myself and my fellow defendants were intending to draw attention to and prevent.

These four simple facts shine a light on a world of evidence you were prevented from considering in this trial, and it is common sense that if you are not allowed to see both sides of the argument, you cannot be sure of guilt. It is also, therefore, obvious that this has not been a fair trial—not by any common-sense definition. Is this how justice should work? You have heard already, you have to be sure to convict. In any UK court, you have to be sure. As a lawyer said in another trial I was involved in, even if you are just a tiny, tiny bit unsure, the law says you have to find the defendant not guilty. This is because a defendant is innocent unless proven guilty. The prosecution has to prove beyond doubt that there was no reasonable excuse in this particular case, as laid down by an act of Parliament.

IN THE CROWN COURT AT SOUTHWARK

HHJ HEHIR

REX

-V-

ROGER HALLAM
DANIEL SHAW
LOUISE LANCASTER
LUCIA WHITTAKER DE ABREU
CRESSIDA GETHIN

FACTS NOT IN DISPUTE

---

1   On 17 December 2020 Her Majesty's Treasury published the New Zero Interim Report which states "Climate change is an existential threat to humanity. Without global action to limit greenhouse gas emissions, the climate will change catastrophically with almost unimaginable consequences for societies across the world". In recognition of the risks the UK became, in 2019, the first major economy to implement a legally binding net zero target.

2   Scientific consensus is that beyond 1.5 degrees celsius warming above pre-industrial levels risks catastrophic and irreversible consequences for humanity which will be irreversible.

3   Over the past five years the global average temperature rise since pre-industrial times has averaged just under 1.3 degrees Celsius. For the 12 months to June 2024 it averaged 1.63 degrees Celsius and is estimated to top 1.5 degrees Celsius permanently before 2030.

4   In October 2022 the UK Government opened the 33rd licensing round to allow oil and gas companies to explore for more fossil fuels in the North Sea.

The Four Facts

Of course, the judge has instructed you that the facts about the massive harm created by the emission of greenhouse gases are not relevant in this case. But you should note that he has also conceded that, to quote, "The facts are as you find them." You have to interpret those facts, you have to make sense of them, you have to decide what it means—what a fact, an agreed fact, means when it says we face a catastrophe, catastrophic consequences as a country if emissions are not stopped. We face some cold, hard facts in this trial. The judge might want to avoid them, but you make up your own mind. The facts as you find them—this is your job. This is what the law on jury equity means. You have to make up your own mind. You're independent in this court. It is a great responsibility, and given the agreed facts, your responsibility goes well beyond this courtroom.

This charge, as you know, has two aspects: the charge of conspiracy—conspiracy to cause a public nuisance—and then also the charge of public nuisance itself, public nuisance without a reasonable excuse. I will deal first with the conspiracy element. I was not part of any conspiracy. I have said so on oath. I was not party to any agreement to go to the countries. I did not go to any organising meetings. I received no paperwork or documentation. I took no trips, and I did none of the organising. It is not up to me to prove to you these things, but for the prosecution to prove that I did receive paperwork, go to meetings, etcetera. And you all have noticed that they have provided no evidence—no evidence at all. And the reason is because there is no evidence. I was simply not involved.

The prosecution says they have powerful evidence. Notice they didn't say conclusive or overwhelming evidence. What they mean to say is there's some evidence. Some evidence is not proof. You cannot be sure, and if you cannot be sure, then you cannot convict, because you don't have the documents, the equipment, the reports of meetings—there was none.

Is There A Time Limit Or Not?
A Letter To Celebrities

Nothing will change without you making ultimatums. Celebrities need to engage in civil disobedience to spark a revolution.

(Aug 12, 2024)

A lobster in a pot,
Dreaming a while
Under the summer moon

– Basho

Note: *This article was transcribed from a prison phoneline. You can listen to the audio version here.*

First things first, thank you so much for speaking outside the court last week and for publicising the appalling mockery of the trial. And thanks to all the others who have spoken up as well. But please remember, this situation is not about us, the #WholeTruthFive. The British legal system is institutionally unjust. Young people are kept on remand for two years or more, and many others have no voice in our brutal prison system.

I am writing this open letter to remind you, and those reading, that we are living through the most fucked up moment in human history. It is so easy for us all to follow the well-worn script: protesters fight for justice, get banged up, liberal notables come to the rescue, the fight continues, and then the good guys win. The anti-slavery campaigns, women's vote, trade union and civil rights, and all the rest of it—all good stuff. But to think that this is where we are at in 2024 is to be as willfully deluded as our friend Judge Hehir, for two reasons.

First, the situation is beyond massive; it's global and murderous. The injustice is incalculable: a third of Pakistan underwater, 50 degrees plus in India, millions of refugees streaming out of the Sahel and Central America. The obscenity of destroying our children's lives—it goes on and on and on.

Clive Lewis MP speaking at our rally. It's time to give the Labour Party an ultimatum.

Secondly, and most critically, there is a time limit, is there not? Or are we still pretending? We are on a cliff edge. This movement was founded shortly after the IPCC report in 2018, saying we have a decade left to half emissions. We will go over 2°C in the 2030s: one billion refugees. We have two years left according to the UN, as outlined a few weeks ago. I could give you another hundred data points: methane, glaciers, forest fires, ice, ozone, acidification. We are in a total fucking emergency. As the civil resistance movements have been pointing out since 2018, if this does not stop, our kids will starve. These islands will become uninhabitable. What was that about the AMOC collapse? Minus 10 to 30 degrees temperatures in UK winters—and then it goes on forever.

The last of our worries is getting banged up for a year or two. Please understand your responsibilities. Causes have demands; emergencies have ultimatums. Nothing will change without you making ultimatums. Clive [Lewis] needs to say to the Labour Party, "Cancel the Tory licences, or he will resign." Jenny [Jones] needs to glue herself to the House of Lords benches. Chris [Packham] needs to lead the marchers to sit down and keep doing it until he's put in prison. All of you need to engage in civil disobedience to force the national media to have a two-hour documentary on the starvation of our children.

Chris Packham needs to lead a mass civil disobedience march.

I am tired of the lazy excuses on all of this. Look at XR, Insulate Britain, Just Stop Oil, and now this trial. There is no time for anything else. The liberal classes have to resist; otherwise, we will have fascism. In the 2030s, the next generation will go, "You did what? A march, a speech? A petition? What the fuck?"

But most of all, this is what I have to say: our civilization is built on the delusion of control, the utilitarian privilege of being able to ask, "Will it work?" You're not put in this world to stand by, to hedge, to prevaricate. We are put here to do the right thing in the face of evil. That is why we got dragged out of court for acting upon this truth. Now the baton is being passed to you. For the sake of the next 10,000 generations, don't drop it.

Who's the Real Fanatic?

Conservatives have lost their way. They are jailing the people trying to stop social collapse. In fact, they have become the fanatics.

(Jun 29, 2024)

> **Note:** *This article was transcribed from a prison phoneline and originally published in the far-right newspaper, The Daily Mail, as part of an effort to appeal to a **conservative** revolution.* You can listen to the audio version *here*.

During my sentencing on Thursday, I was called a fanatic by Judge Hehir. This word was published on the front pages of the national press. I think I have a right to respond.

For over two decades, I was a farmer in Wales, supplied my community with vegetables, and lived a family life. I worked very hard, paid my taxes, and was no harm to anyone. I was doing my job, but the politicians were not doing their job. Along with millions of other farmers, I was forced out of business by extreme weather events—months without rain, months with nothing but rain, now yearly events.

Central to my personal belief is the idea of balance. I have the same orientation as many traditional conservatives, such as Edmund Burke. Destroying the weather systems in order to take money is not balanced; it is a crime. My political views are rooted in the philosophy of John Locke, the father of classical liberalism. We should live under the rule of law, and so should the government. The basis of the law is the welfare of the people. If a regime enables the destruction of the lives and livelihoods of the people, then the government is a tyranny, and the people have the right to rebel

Daily  Mail

FRIDAY, JULY 19, 2024   dailymail.co.uk   **£1.10** 85p to subscribers

Secrets & Lives
STARTS PAGE 29

How I learned the hard way why women like Strictly's Zara put up with toxic men

BRYONY GORDON

As five Just Stop Oil activists given record jail terms for M25 chaos...

5 YEARS	4 YEARS	4 YEARS	4 YEARS	4 YEARS
Roger Hallam, 58	Cressida Gethin, 22	Daniel Shaw, 38	Lucia Whittaker De Abreu, 35	Louise Lancaster, 58

JUDGE WHO SPOKE FOR ALL OF US ON ECO-FANATICS

By **Colin Fernandez** Environment Editor

A JUDGE blasted 'fanatic' eco-activists who paralysed the M25 as he handed them record jail terms yesterday.

Jailing Just Stop Oil co-founder Roger Hallam for five years – the longest-ever sentence for a peaceful protest – the

Turn to Page 4

"Each of you crossed the line from concerned campaigner to fanatic... bound neither by the principles of democracy nor the rule of law... heedless of the rights of your fellow citizens"

Front pages of the billionaire-owned, Daily Mail and Daily Telegraph on the trial.

175

No "functioning democracy" enables the obliteration of our way of life for the next hundred thousand years. Or are we living in 1984, where the consequences of going over two degrees by burning new oil and gas is a continuation of the rule of law?

Judge Hehir said in the trial that these effects of what he called "climate change" were "neither here nor there." He was given 200 pages of scientific evidence but refused to change his mind. Furthermore, he refused to let the jury see that evidence so they could make up their minds. He merrily declared that the human race going to "a fiery end" was not relevant in a British court.

Can you imagine if he had said that about the 50 million killed in World War II? If he had said that about the millions murdered by the Nazis? John Locke was clear: any government that plans the death of its citizens is breaking the law, in the past, in the present, and in the future. No exceptions.

When our children read about what Judge Hehir said a decade from now, they will have one word to describe him: fanatic.

I'm smiling - are you?

Integrity vs Expediency: The Climate Trial

As Martin Luther King said, moderates are our biggest obstacle to freedom. Here's how they enabled my prison sentence.

(Jul 24, 2024)

Note: *This article was transcribed from a prison phoneline. You can listen to it here.*

Yesterday, myself and four others were imprisoned for four to five years. The conservatives and the liberals have got it all sewn up. The narrative is set: for the conservatives, it's job well done. For the liberals, it's another chance to go through the motions of an "injustice trial". But this is not about five nice white middle-class people being banged up for "protest" on "the climate." It's about a few million not-so-nice white people deciding to have a few hundred million brown people die. Just for starters.

This is not about "climate change"; we agree with the judge on that. It is about murder. At scale. Forever. And that is a bad thing, a very bad thing, an evil thing. When the United Nations recently said we have two years to save the world, that they are not being "melodramatic," that economies will be "devastated," they mean it. Not in some distant future. In the next 10 to 20 years, that's what 1,000 public statements have said. It's what 10,000 peer-reviewed papers have said. It's coming. It is what it is. At some point, you'll be stepping over body parts on the way to work, going "well, you know."

What do you think David Attenborough means when he said "we face the end of civilization"? A picnic? What do you think 1 billion refugees at 2 degrees Celsius, 20 times the number at the end of World War II, looks like? Really?

Conservatives are the bad guys and liberals are the bad guys who pretend to be good guys, and the latter are the worse, which is why historically they are held in more contempt. They knew but they did nothing. As Martin Luther King said in a letter from a Birmingham jail, it's the moderates that repress, distract, sabotage the resistance to injustice. They are the main problem. Of course, the Carbon State and its functionaries are going to put people in prison for years if they come up with a resistance plan proportionate to the level of criminality we objectively face. If they are willing to have a few hundred million black people starve to death, then why be surprised that they pervert the course of justice?

I must make two honest confessions to you, my Christian and Jewish brothers. First, I must confess that over the last few years I have been gravely disappointed with the white moderate. I have almost reached the regrettable conclusion that the Negro's great stumbling block in the stride toward freedom is not the White Citizen's Council-er or the Klu Klux Klanner, but the white moderate who is more devoted to "order" than to justice; who prefers a negative peace which is the absence of tension to a postive peace which is the presence of justice; who constantly says "I agree with you in the goal you seek, but I can't agree with your methods of direct action;" who paternalistically feels that he can set the time-table for another man's freedom; who lives by the myth of time and who constantly advises the Negro to wait until a "more convenient season." Shallow understanding from people of goodwill is more frustrating than absolute misunderstanding from people of ill will. Lukewarm acceptance is much more bewildering than outright rejection.

Martin Luther King's Original Letter from A Birmingham Jail

This trial was not about "the right to protest." It is not about "a cause," "an issue." It's civil resistance against the biggest death project in human history, the greatest ever act of criminality. This trial was an experiment with the truth, as Gandhi called it. We were not trying to win, we were trying to tell the truth as if the truth was real, as if this slaughter, starvation, and rape is real. It was integrity, not expediency. So obviously, we spoke that truth. Obviously, we got interrupted. Obviously, we continue to speak even when the judges shouted at us to stop, had us dragged from the dock, banged up in jail. This is what evil looks like. It's what it does. And it's just the beginning.

Integrity at the present time is resistance, nothing more, nothing less. It's the opposite logic to expediency. Expediency is trying to have your cake and eat it. To maintain your privilege and status while appearing to do good. Expediency is to write an article about the trial but not glue your hands to your editor's desk. Expediency is to call for justice but not to challenge to judge because it will do in your career. Expediency is to lead a march but not have anyone sit down and be arrested. Expediency is doing everything that looks good but does no good. Expediency is betrayal. Betrayal of your family, your country, this world, but also of yourself—that temporary spark of consciousness in the void of eternity. Consciousness is truth, beauty, love. When you're on your deathbed, you will not be thinking about your career, the stuff you had; you'll be thinking about whether you became what you know yourself to be, a soul.

Integrity is a hard path. Your ego has to burn in a fire that destroys its desire to control. Integrity is humiliation, failure, being forgotten. As the greatest soul of the 20th century, Simone Weil, said: when you have an important decision to make, choose the most costly option. We might add, "in the 2020s," because if you don't, the cost will be far greater. In a week or two, all this trial business will be old news. All that will be left for us is the dual brutality of a British prison. Today I had boiled rice with fried rice. Yesterday, boiled rice with pasta. A few days ago, a note came through that cell door: two paragraphs, a guy down the row had killed himself. The new inmates bang on their iron doors all night, yelling, caged and enraged.

As this article wrote itself in my head, I cried. Not tears of self-pity, not of anger, but of determination. I know who I am. I know what I am doing, and that's why I'm Britain's most influential climate campaigner, as they like to call me. Take note: in the end times, integrity trumps expediency. I'm smiling. Are you?

Law - What is it good for?

Roger is in prison and facing the longest sentence in UK history for peaceful protest. It's time to help.

(Jul 17, 2024)

Tomorrow, Roger, along with the rest of the #WholeTruthFive (WTF), will be sentenced to what will likely be the longest prison sentences for peaceful protest in UK history. Their "crime"? Participating in a Zoom call to discuss proportionate disruptive action against catastrophic new fossil fuel licences. WTF!

The UN has slammed the trial for persecution and the harsh penalty of its defendants. During the trial, the five weren't allowed to use the science of how putting carbon in the atmosphere will kill millions (the whole truth) as evidence in our defence for disruption of the M25, one of the UK's most polluting motorways.

The justice system is fucked.

Stand up against it, before you're next.

The defendants alongside Defend Our Juries activists and the UN Special Rappoteur on Enviromental Defenders, Michel Forst, who slammed the controversial trial.

Jailed for Telling The Truth

**Roger was imprisoned this week for telling the truth in a British court. He refused to leave the witness box after the judge ordered him to stop giving evidence of climate breakdown.
The following is his statement.**

(Jul 13, 2024)

I have been imprisoned, along with three of my co-defendants, by Judge Christopher Hehir.

As you may know, I have been in court charged with 'conspiring to cause a public nuisance' for making a speech on civil disobedience before Just Stop Oil blocked the M25. During the trial, I stuck to my oath to tell the jury "the whole truth". It is their fundamental right to hear all the evidence.

In response, the judge stated that "whether or not we are facing the end of the world is neither here nor there" and that humanity "coming to a fiery end" was irrelevant. He then ordered me to be forcibly dragged out of the court by the police and remanded to prison. This is the indignity of a British courtroom.

The corruption of our judges by the carbon state has crossed a line in the sand. This is an opportunity, and an obligation, to act. We only have a limited amount of time to halt the unimaginable horrors of climate and social collapse - and to save our democracy.

I call on people - you reading this - to come to the court with placards to make clear the jury has a right to the whole truth and nothing but the truth.

Act in solidarity not just with us, but with the billions of people whose lives and livelihoods will be destroyed unless we stop this greatest of all crimes.

In truth, there is simply nothing more important than to stop this. Is there?

Untitled (Fallen Angel), acrylic and oilstick on canvas. Jean-Michel Basquiat (1981)

I'm out of prison and mad as hell

(September 24, 2023)

I got out of prison about a week ago after nearly four months. As you may know, I was imprisoned for making a 20 minute speech on zoom. That was it. I was on remand. I have yet to be convicted of any crime. I have not committed any crime. I am applying to the court to have the charge of "conspiracy to cause a public nuisance" dismissed. I did not conspire – I engaged in an act of speech. And what happened – the blocking of motorways by getting onto bridges, was not a public nuisance by any stretch of the imagination. Given what we face. Given what this society has done. What is now locked in.

If the case is dismissed, I will look into suing for wrongful imprisonment. If they don't dismiss it, I will be on the (nonviolent) war path. I know who I am and I know my rights. I will not stand by and have them violated. Not because of the harm it does to me personally. That is the last of my concerns. They can put me in prison whenever they want and for as long as they want.

No, I am doing it because people died so that our generation and future generations could enjoy the freedoms and liberties of this country, and the last thing I am going to do is standby and betray those who made the ultimate sacrifice so I can live. That's about as basic as it gets.

I'm mad as hell and that's a joyful place to be. Because only when you are as mad as hell – facing what we face, can you ever feel free in the real sense of the word. Only when you are mad as hell – facing what we face, can you have any respect for yourself. Only by being mad as hell then, can you be who you truly are. If you don't get this in 2023, with respect, you simply don't get what it means to live a life. My mother said to me "Life is for Living" – meaning, it is what is it and you stand up and face it.

My last cellmate in prison was supposed to be there for two weeks. They messed up his paperwork and he didn't get out for six weeks. What did he do? He got on with it. He wrote applications each day to various prison departments – all that stuff. When his girlfriend told him on the phone for the first time that she loved him, he was over the moon. He jumped off the top bunk in glee. He had been in a terrible car accident and should have died. Two months in a coma. He had a big dent on his head. He could only see out of one eye. He didn't look that good. Had a drink problem and all the rest of it. But he was living his life. He's my hero.

Remember this is not a drill, as has been said. You're reading this post. Fine then go and live your life – meaning doing whatever it takes to stop the apocalypse we face. Even if it kills you. Really – that's what living a life means at this present moment. Just as it did in previous times of total crisis.

I don't know how I will end up and don't much care anymore. I've seen over the mountain top, though no doubt often I will forget that I have. I've had a glimpse of what it is to be truly human. I will stride onwards. And it's in this striding I will serve myself, humanity, and this world. Because at the end of the day it's all one thing.

There's not much more that can be said than that.

Thanks so much for all the supportive letters and postcards. See you on the streets.

Image: *Angel of the Revelation* (Book of Revelation, chapter 10). Watercolour, pen and black ink, brush and wash, over traces of graphite. William Blake (1803-5)

After Despair: Awakening to the Revolution in 2023

(January 13, 2023)

I've been to prison a good few times (and I'm still here), and when I get out people go "was it really bad in there?". I always feel like saying, "yes, it was, for the same reason it's really bad being out of prison". Meaning just about every day, I find myself having feelings of utter despair, about the horror and terror that is coming down the road for humanity. Sometimes I can't help myself, prodding this monster. Sometimes, the monster jumps out of me and consumes me. I am helpless to resist. There is no reasoning out of the abyss. It seems to come and go of its own accord. When it is particularly bad, all I can do is stare into space. My partner gives me a massage and I sleep. And often I'm blessed with it disappearing by the morning. Sometimes though, it can go on for weeks.

It is as if I travel into the future. A vision happens. There is a cascade and compounding of desolations, extreme weather beyond any experience, the devastation of crops and cities, slaughter and rape, mental breakdowns, suicide, and infanticide. Coming and going and coming back year after year, decade after decade, until there is only a deathly silence, a final endless void. a final endless void.

I feel embarrassed writing this like I should at this point put down a few facts and figures "to support my viewpoint". But I and others have been doing this meticulously for years now, to no avail. This is not a viewpoint but a world of emotional reality. It is the greatest tragedy of the human condition, that denial, the avoidance of all of this, is so functional. We cannot socially operate without it. And at the same time, it is so appallingly dysfunctional in that it is denial that prevents us from stopping the horrors from happening. And it is the greatest paradox of human existence, that it is the very experience of going through the hell of despair that gives individuals the miraculous strength to stand up to injustice and achieve the impossible. To provide a pathway, if there is going to be one, out of a catastrophe we are entering.

This is the perennial story, the dark night of the soul, the hero's journey. In our society, all this is avoided and ignored or kept behind the screen. Sanitised and mystified. But actually, nothing is more real than this, or as messy. Some seem to avoid it, others fall into the pit and never get out, and others come out and then fall back in. There is no certainty. But the general path is clear enough. It has been told countless times in our histories and in the great stories of our cultures.

In modern language, we might call this the death of the ego. In the present context, I would call it the death of "the reformist ego". The moment of revelation here is that we are not actually despairing of the world, but rather despairing of our "selves".

A certain idea of the "self", we have been given. Waves of agony, literally burn this "self" to death. This is the "self" which desires to do our bit, that wants to be nice, to go with the crowd. The "self" that hopes it will be all okay. That it will not need to sacrifice its status, its assets, its relationship; family, career, or property. It's all bollocks. The raging fire of self-contempt shatters the lies we live by.

This process is the exact opposite of the perversion of "regenerative culture" into self-love - the idealisation of the self. It's the opposite of "deep adaptation", in so much as this phrase refers to the escape into a privileged individualism and politicisation.

It is the burning flame of social revolution that will bring down and take over the carbon regimes. This is utterly un-abstract. Is the concreteness of a collapse of the global neoliberal edifice. Its power holders will beg us to take over from the chaos they have created through decades of lies and deceit.

I remember a senior diplomat from COP privately pleading with XR co-founders to come and close down the Chile conference. They are all zombies and they know it. 30 years of false promises.

This then is the awakening to the revolution. The explosion of a collective desire for redemption and salvation. Only those that have been through the agony of the dark valley can provide the leadership to bring the people to the mountaintop. So goes the universal myth. It becomes real, all too real, in the enactment of new sociability, and collectivity, combined with the miracle of innovation and organisation to rapidly decarbonise and engage in earth repair. Only a revolution can do this. And only a revolution will be able to save us.

There is everything to play for. In fact, the story has only just begun. Insulate Britain, Just Stop Oil, and the other A22 Network campaigns around the Western world, are just the prelude. Our task now is to revive the vision of the world we'll be calling out for, as an alternative to the universal death proposed by performative nihilism and pathological fascism.

It is difficult to write this post and I am sure I have not done justice to the great depths and variety of experiences involved in these processes. I hope, along with others, to develop the practical details of this revolution in the coming year. And I will write about this in Facebook posts and speak about it in my new series of podcasts/videos "Designing the Revolution".

But it is vital that we collectively acknowledge that there is no reason without passion - as our radical traditions make clear, perspectives rediscovered by modern psychology. Self-love must be sacrificed in order for us to move into service to the other, to the universal. The reformist ego must be burned to death in order for us to awaken to the revolution. We need to get our existential ducks in a row. And then... Onward!

Image: *The Prisoner.* Nikolaj Alexandrowitsch Jaroschenko. Oil on canvas, 1878

English Gulag

(December 29, 2022)

About nine weeks ago, I did a 20-minute speech on the crisis. Just like the 100+ public talks I have done over the past year.

Destroying sovereign states, creating billions of refugees, shitting on our children's inheritance, and wiping out what's left of the natural world justifies concrete acts of civil disobedience. Obviously. Scarlet Howes from The Sun newspaper made a recording, she gave it to the police and made a deal. The Sun would get to film my arrest when the police came into my flat. I was not there when it happened. They took my laptop and other stuff like they do every few months. They changed the locks and left a note for me to come to the police station.

They arrested me. 36 hours in a police cell. The interview by the police involved 200 Questions. 200 times I said "No comment". Not one question on me talking about James Hansen, ex-head of NASA, and his colleagues' latest memo confirming the world will pass 1.5 degrees in 2024. Not one question on me saying that the worst episode of suffering and injustice in human history more than justifies nonviolent disruption under English Common Law. Right of Necessity. Some evidence cannot be collected. Police procedure has become politicised. You might call it that. Five minutes in court.

The magistrate put me on remand. For a speech. "Anything else?". She does not look at me. "Chop chop, time is up". No time to lose, at least when it's her time.

In the prison cell, it feels like 90 degrees. My Romanian cellmate and I are down to our underpants. Outside it's the hottest November on record and the heating is on full blast. There's no water in the cell. For three days, the toilet won't flush. The plumbing's down on the whole wing.

I talked to a prison officer who seemed important. "There's no water in the cell, I suggest you make some phone calls". It's only my second day. I'm trying to be polite.

We're soon down to our last half-litre of water. Water's "on the way". Me and the Romanian look at each other. He doesn't speak English, but we're both thinking "This is when we start to panic". That night, shit is being thrown out of the cell windows. Literally.

I got moved to another cell with a career criminal, Eddie - he's had 19 years inside. He tells me the latest, a young guy of 21 has just hanged himself. After he'd had breakfast, apparently. "Poor fucker". And he's not the only one. Eddie's been here for months and he's settled in. Then he's given one hour's notice that he's moving to another prison. He's not happy, but he knows there's nothing he can do.

My next cellmate is a drug addict. Three times a day, the smell of burning paper, smoking out the cell. It's spice and he's told me the deal. It gets up my nose. I'm watching England play and in five minutes I'm on my back on the bed. I can't move my body, I can see but I can't process what I'm seeing. For the next three hours, I'm thinking "Am I going to die?". If I do then no one's going to come. If you press the "emergency button" it's an hour before it's answered. "Emergency" here means emergency as in "Climate Emergency" as declared by the British Parliament three years ago. Some things are connected.

My cellmate breathes so heavily that I cannot think. When he coughs it's like he's going to die. I tell the prison officer "He's going to die". He says he'll sort it out, as they do. Later that day the prison officer sticks his head around the door. "You alright?". That's it. On telly, it's about Chris Kaba, shot dead by the police. It's June 2023 before they find out what happened. Years before they get any justice. Black Lives... right. You "alright", Chris? A week later, my cellmate finally gets moved.

I'm supposed to be able to get three visits a week. After hours of waiting on the phone, and a dozen rejections, finally my partner gets to see me on week six. The same week, I got my first visit from a solicitor. No, they don't have a full transcript of my speech. No, they don't think there's any other evidence. No, they don't advise me to appeal to end my remand. "Because it's you". I'm a co-founder of Extinction Rebellion. Did the design for Insulate Britain. The tabloids say I'm the leader of Just Stop Oil. We all have to make a living, right? But unlike the tabloids, it's not a living selling lies.

The hottest ever November turns into the coldest December since... whenever. The slits in my window won't close. It's -7 outside. I get chapped lips, chilblains, and aching legs. I can't sleep, I'm too tired. The prison officer won't give me any tape to close over the gaps between the slits. "No, you can't take your coat back into the cell after getting to court". Finally, 2 workmen turn up. "It says you've got hypothermia". They don't blink. They get a big pair of pliers and twist the dial with two hands. The slits close. They'd gotten stuck.

Really hot, really cold. Freak weather? No, it's weather-blocking, stupid. 50% of the ice in the Arctic has melted. The cold air slips into the hot, the hot into the cold - winds slow down. It all started 15 years ago, I was a farmer then. 7 weeks of rain every day from June for 2 years running - I lost all my summer crops. Weeks of -15 degrees - I lost all my winter crops. I'm not a farmer anymore, but I'm still here. Others committed suicide. But it doesn't get on the news.

When all the ice is melted in the summer (in the coming years, according to peer-reviewed papers) it'll get 1000 times worse and billions will starve. Then it will get in the news. (Adam, this is the bit where the reader doesn't "Look Up", right?).

Yesterday the heating broke down. Yes, there is a pattern here. As I write this, I've got a t-shirt tied around my head and a blanket over my shoulders and legs. I put my hands down my trousers between writing paragraphs.

The English gulag.

On telly is the cost of living crisis. Kids are sitting in front of the TV covered in blankets, the heating only goes on for 20 minutes a day. Nurses are going on strike. In A&E's waiting rooms, patients are sitting on the floor while their children are in pain in their arms. In schools, the roofs are leaking and sewage comes up into the corridors.

The English gulag.

Do you think I'm talking about prison? The joke's on you. This country is a gulag. A million times a day people give each other that look of "This place is so fucked". It's just that we in prison are on the worst wing. But your cell doors are locked too. It's just that your cells are bigger than ours.

I went for an "education interview". A prisoner says "Aren't you one of them protesters? Well, tell them, when you get out, that three prisoners have committed suicide in here. Tell them that.". He's a caged Tiger, his eyes say more than his lips.

I've just finished Jeffrey Parker's "Global Crisis: War, Climate Change and Catastrophe in the Seventeenth Century". In the 1600s the world's population fell by a third. Due to the climate-war-disease complex, it's all one thing. Slaughter-rape-suicide-desolation. "Our descendants will never believe what miseries we suffered". Except we will because it's about to happen again. Except for this time, it'll go on forever. Co2 is being put into the air 8 to 30 times faster than at any time in the earth's history.

The rumours are coming down the line. Except they're not rumours, they're facts. As they discovered in the Warsaw ghetto in 1943, they are going to have us die unless we stop them. The atmosphere is a gas chamber. It literally is a gas chamber. Except this gas chamber covers the whole world and there's no escape. A few years back when I wanted to say this in public, my PR advisor threatened to resign if I did so. The most important truths can never be allowed to be spoken.

Merry Christmas, Scarlet Howes, and all the best for the new year. A few years ago, when I started my research at King's College London, my supervisor told me I was the best PhD candidate he'd ever had. A few weeks later, he was already getting annoyed with me. He said if I continued to speak the truth, I would make myself "irrelevant". He didn't say exactly that. They never do, do they? But that's what he meant. Three years later, I started Extinction Rebellion. The biggest global climate influencer of 2019. 200,000 people joined in six months.

So listen to me. This is important. To all those caged tigers reading this. Truth is the most beautiful thing in this world. It's irrelevant whether it's "irrelevant". Acting the truth every day. That's what we exist to do. Nothing more. Nothing less.

I'm out of Prison!

(October 19, 2019)

'I found prison in many ways quite agreeable. I had no engagements, no difficult decisions to make, no fear of callers, no interruptions to my work. I read enormously; I wrote a book, "Introduction to Mathematical Philosophy"... and began the work for "Analysis of Mind" '

– Bertrand Russell about his 6 months in prison

Hello lovely people,

Yes, I am finally out of Wormwood Scrubs after six weeks. (They have a nice front door as you can see). Thank you so much to all the kind people who emailed me messages. It was fine, I was fine and if anything was annoying, it was sitting there thinking about why so many people make up excuses for not taking this step which is now so necessary. I know I'm a bit weird

– sitting on a bed all day reading biographies of Gandhi, having my food made for me, is pretty much as good as it gets as far as I am concerned. But really going to prison is not the end of the world.

Here then are a few "obviously's":

1. Obviously going to prison is not for everyone. If your mum is about to die, if you are traumatised, if you are claustrophobic, then it's not for you.

2. At the same time obviously thousands of people from all backgrounds can do it and the excuses are just due to prejudices and unfounded fears.

3. Obviously you want to be sorted before you do it. It's a project, like moving house, going to another country for a time, or changing a job. It's a big thing but it's just another thing.

4. Obviously going to jail is not going change the world just like that. Like pressing a button and everything is sorted.

5. At the same time obviously going to prison is a massive headache materially and politically for the authorities and once hundreds and then thousands of people do it in an organised and coordinated way, then in my scholarly opinion real policy changes are going to happen. It is the single biggest act of radical political and nonviolent effectiveness any social movement can decide upon.

I think on the last count something like 1800 people in Extinction Rebellion are up for it. So I will be spending a lot of my time helping to get actions organised which enable these people to walk their talk. If you are reading this, then I am sure you know it's all systems go, the time is now, and all the rest of it on the Ecological Catastrophe. So get in touch with XR, book some time off, and make up your book list.

And thanks again for the support – I really appreciated it.

Additional note:

Hi – Thanks for all the comments. I am very sorry some people are upset by what I have written. Discussing prison is very difficult because it is a terrible experience for many people and it is very definitely not for everyone which is what I am trying to say in the post. However, it is a debate which has to be had at a time when dramatic action is so necessary. Social breakdown is going to happen unless we stop governments allowing the emissions of carbon which is leading to the destruction of the climate and ecological systems. I have been in prison five times and I am speaking from my own experience and the views are my own, expressed on my own personal facebook page. I would appreciate it if the responses are respectful. Thanks.

Final Comments

Beyond the Big Lie: The Return of America's Class Struggle

(November 12, 2024)

America's 2024 election reveals a brutal truth: behind the masks of progressivism and populism, the battle is still about class—and the stakes have never been higher.

> *"It is better to be less understandable, less pleasing, less moving, than to speak what is not true and not just."*
>
> – St. Augustine

The ancient Persian religion Zoroastrianism was founded on rejecting "The Lie"—not merely because lying is wrong but because it violates humanity and is a crime against the cosmos. Truth was seen as holy, while lying was a path to ruin.

The truth is that America, like Britain, is now run by the rich, for the rich. To say otherwise is a lie. And for the wealthy to rob the poor is unjust. It always was and always will be. This is a fundamental principle. To steal from the poor is one thing, a material wrong. But to do so while pretending otherwise is something else entirely. That act doesn't just rob the body but corrupts the soul — a far greater crime.

This is the crime of the Democratic Party. For decades, it has talked the talk while betraying America's working class. Claiming the "progressive" mantle, it has actively transferred wealth from the poor to the rich, destroying communities in favour of quarterly returns. We all know this. When you lie cynically, you corrupt not only yourself but your audience as well. If you make people gaze into the abyss long enough, that abyss will gaze back at them. From that abyss comes fascism. When elites — those we're told to respect — speak nonsense, people will take that as permission to do the same, believing that a billionaire can deliver justice. Voting for Trump is a people's defiant response to those who have insulted their souls with a 'Big Lie' year after year after year. Being subjected to such deceit inevitably breeds a desire for self-destruction.

Power, especially the power of Capital, has one playbook: divide and rule. For decades, its "progressive" representatives have told us it's about race, gender, and ethnicity. That's a lie. At its core, it's about class. Yes, the liberation of specific oppressed groups is essential, but only within the context of the larger story: the oppression of all groups by the rich. Progressivism in America has become rich people telling the poor who's more oppressed than whom. It's vile nonsense; it's a Big Lie.

The results of the 2024 election demonstrate that the game is up. People have stopped thinking of themselves solely by identity labels like "women," "Black," or "Hispanic." Instead, they see themselves as "the people" — people who can't afford basic groceries. And with no candidate truly standing up for their class interests, they've turned to the next best option: someone who claims he will. Because, yes, he's a fascist, but he's not part of the managerial elite — that administrative class of university-educated people who talk the talk but in reality, micro-manage the working class in service to a state that serves Capital. Who would vote for that?

In the past four years, the Democrats have overseen record-high oil and gas production, and Trump will continue this suicidal course. Soon, the elites will have made Florida uninhabitable, forced evacuations of the American Southwest, and caused millions of refugees to flood across borders. To claim otherwise is the greatest Big Lie of all. People will then face a fundamental choice: to embrace a fascism that sacrifices the vulnerable, an anti-Christian civilisation, or to choose a socialism that extends charity to the vulnerable, a Christian civilisation.

So what's the plan? The plan is to stop lying. America needs a political party of the people, by the people, for the people. That will never be the Democrats. Their lies have burnt their bridges. The answer lies in a genuine democratic socialist force, but one that moves beyond stale language. People don't just want "equality" — they want family, community, and nation, all of which are being destroyed by capitalism. They want freedom, dignity, and the chance to create and flourish, which requires ending the neoliberal stranglehold. They don't want to be ruled; they want to govern themselves, as the American experiment was supposed to promise. Not another oligarchy in new clothes, but a movement that uses the people's conservative values to stop the rich from robbing them. To speak the truth, even if it's less understandable, less pleasing, or less moving — because it's the truth.

If a real choice is available, a political force for the latter must start now. To make America truly great, not by superficial strength, but by being true and just — and meaning it.

How to Stop Fascism

(November 06, 2024)

It's through connection, not ideology, that we will win the revolution.

One word. **Proximity.** Let me explain.

In a now largely forgotten book published in 1989 called **Modernity and the Holocaust**, Polish-Jewish sociologist Zygmunt Bauman examined the overwhelming evidence for why the Holocaust happened. It required a gang of political adventurers who seized control of the German state, wielding its immense organisational and industrial capacity. But this still leaves the question of **how** they managed it. In a key section titled **Social Proximity and Moral Responsibility,** Bauman argues that the decisive factor wasn't ideology but the creation of psychological and physical distance between Jewish people and the rest of the population.

In 1933 the Nazis called for a boycott of Jewish businesses. It failed because too many non-Jewish Germans personally knew their local Jewish shop owners and felt a moral responsibility not to harm them. The Nazis learnt from this and, over the next decade, progressively reduced the proximity between Jews and other Germans. Jewish people were forced to wear identifying badges, then moved into camps, and ultimately, when they were murdered, the separation had become so vast that most people felt no moral responsibility to intervene.

To prevent fascism, we must do the opposite. A pro-social movement of political adventurers must take control of the state and progressively increase proximity. Proximity—not "economic growth"—is the key determinant of a flourishing civilisation. It's what people truly crave: to connect, to sit down in small groups and talk. We care about "politics" and "self-interest" only to the extent that they foster this connection. Love— mutual care and recognition—is humanity's foundational need. This truth spans the social sciences and is echoed in the wisdom traditions.

House of The People Gathering

Let me give a few examples. In my award-winning research at King's College London I showed that if people sit in small circles to discuss a social issue (with biscuits on the table!) for most of a public meeting, 80% leave feeling empowered. In contrast, only 20% feel empowered after a conventional meeting with a series of speakers and no small group discussion. Research shows most people initially attend campaign meetings not for political reasons, but because a friend invited them or they seek human connection. A Harvard study on negotiation found the single biggest predictor of success is whether the other party personally likes you. The early Christian church, one of the most successful movements in history, didn't convert people through doctrinal persuasion but by fostering friendships. The evidence is overwhelming. "It's absolutely fantastic," one trade union leader told me after restructuring his events around small group discussions.

Progressives and the Left fail repeatedly because they are wedded to an Enlightenment secular religion that assumes people respond to ideas over emotions and ideology over connection. They're embarrassingly mistaken.

Fascist men, for instance, often abandon their views after forming personal relationships—such as getting a girlfriend. There are countless stories of individuals entering far-right spaces, listening, building personal connections, and subsequently helping others to leave those spaces. During the recent English riots a group of Muslims, faced with an angry crowd outside their mosque, offered food and listened to people's concerns. Conversations ensued, tensions eased, and constructive dialogue began.

I've given over 200 public talks on the climate crisis, often with right-wing conspiracy theorists turning up to disrupt. I never argue with them but thank them for sharing their views and invite them to discuss afterwards. I take them aside in a small group, summarise their questions as they speak, and they continue talking. It works wonders. One of the main right-wing conspiracy theorists in the UK even shook my hand afterwards. Proximity can build a connection in minutes.

Of course this doesn't work every time, and ideas have their place. But as Carl Rogers, one of the most influential psychologists of the last century discovered, listening with "unconditional positive regard" is the single most effective way to promote personal healing and growth. Rogers sparked a counselling revolution, transforming therapy. We need a similar revolution to transform politics. Change won't come through conventional politics but by circumventing it. We must create proximity directly—through door-knocking and local assemblies organised around small group listening.

This is the next political revolution. I'm convinced that new movement parties could rise from zero to winning local and national elections in six months with these methods. We've already seen similar success with new parties in recent years. But it needs perfecting. A national project would require around £1 million to get started. Anyone with money out there would see a far better return by investing in this method to counter the far-right threat, rather than wasting funds on conventional political strategies.

We can either build a new civilisation based on closeness and connection or let humanity destroy itself by ignoring how we actually function. Getting this right, right now, is of utmost importance. You know exactly what I mean.

Hothouse Holocaust: Awaiting The Glory of Resistance

(October 29, 2024)

Humanity stands in the darkest valley — caught between denial and panic, waiting for the catastrophe that will finally break through our apathy.

This article was recorded down a prison phoneline. You can listen to the original audio with subtitles here.

On our present path, civilisation as we know it will disappear. Only if we meet current commitments — net zero by 2050 — perhaps some form of humanity will survive, managing the challenges of continuing extreme weather events, ice loss, and sea level and temperature rises.

Sir David King, former Chief Scientific Adviser to the UK Government, 2024.

What strikes you about the title *Hothouse Holocaust*? Well, obviously it's the "H" that starts each word, the way it rolls off the tongue. It's not really about what it's saying; that somehow drifts into the background, some vague unpleasantness. The H's — that's the thing. Human beings, eh? We're a strange bunch. I have this sneaking feeling that when it becomes official, and someone very important finally comes on telly to tell us, *"Unfortunately, it is true. After all, we have left it too late. Climate collapse is now locked in; billions will now die"* — when this happens, there will be a not-insignificant number of viewers who will be more interested in his tie being out of place or that there's something wrong with his hair. Anything but what he is saying. Human beings...

We've heard it all before anyway. It's been in the air for a long time; those not enmeshed in the mania of denial can smell it. It's not good. It's not going to end well. A week or so before writing this, it came out that the carbon sinks are collapsing. Last year they <u>failed to absorb more carbon than they gave out</u> — the forests, the soils, the oceans. It could not be worse news.

I did a tweet. I wrote "fuck" 4,657 times. Yes, I was wondering too whether that is a world record — at least from someone in prison.

The billionaire Elon Musk has given me £700 from his fortune for my efforts, so that's something, I suppose.

It got 35 million views, but no one's on the streets. That sums up our present moment. I don't want to sound overdramatic, but isn't there someone out there who agrees with me? We are living in the most morally decrepit time in human history — this unbearably surreal moment between reasonable denial and panic-stricken action; the darkest valley of spiritual death, where we can no longer deny but not yet act. We can click on the tweet by the tens of millions, but going to the street? *"Make me good, but not yet."* Each day is just a waiting, waiting for that one-in-a-million-chance climate event that is now going to happen once a decade for the next hundred thousand years — which finally breaks through, that kills tens of millions of Black people or a few hundred thousand white people (after all, racism has not gone away, right? Let's at least be honest about that).

Waiting, waiting, waiting, and then it happens.

The phone rings. It's some rich person in a mad panic; he's never donated money in the past, but now he wants to give us millions. Each day of waiting for that phone call means more uncountable, unique lives are condemned to slow, agonizing starvation. I play with variations on the most human of reasonable responses: *"You fucking idiot, you should have given it 10 years ago. Now it's too fucking late."*

Why We're Going Extinct - 5000 Fucks, 30 Million Views, 8 Billion Dead

(October 22, 2024)

My tweet on collapsing carbon sinks has gone viral, racking up 30 million views. But what does 5,000 'Fucks' mean for our path to extinction? 5,000 Fucks, 30 Million Views, 8 Billion Dead

This article was recorded down a prison phoneline.

We are not going extinct because of the climate crisis.

We're going extinct because we can't think straight.

Let me give you a simple example.

What, dear human being reading this, is worse: a child trips over a railway line and cuts her knee, or a child is tied to the railway line and a train is coming down the track?

Okay, yeah, it's the second situation. Because although the harm has not yet been caused, it is going to happen, and the kid will die.

So far, sound good? Very clever of you. So try this one: what is worse—Auschwitz, the Nazi concentration camp, or a thousand Auschwitzes, the thousand concentration camps coming down the line due to endless global chaos caused by elites allowing the continued emission of carbon?

Yep, you got it right—it's obviously the Nazi horror, right? Because that happened, and 1,000 camps coming down the line have not happened yet, we don't need to worry about that, right?

Burn, baby, burn.

We are going extinct because of your "right answer."

Maybe a few of you feel (obviously, this is subjective) that 1,000 is greater than 1, so you don't want us to go extinct. In which case, for the small percentage of you who have the attention span to get this far down this tweet/article, you might want to know what we need to do.

Well, as one of the world's top campaigners, sitting in a cell after being given a five-year sentence for suggesting we do something effective about going extinct, I have a few ideas I will humbly share with you.

First, humility will get you nowhere. The only thing that gets attention is drama, and without attention, you don't even get past the starting line.

Don't write "scientists are concerned." Don't say you're really upset. Don't put in a one-off "fuck." No—put "fuck" 4,568 times. Then you might get 30 million views rather than 1,000.

This is what I call the Larry Kramer Law. If you're not up to speed on the guy who should be considered the greatest campaigner of the last century, you should know that he did not hold back. He turned around the disgusting neglect of gay people dying of AIDS in hospital corridors during the 1980s in just six months.

He didn't ask nicely for people to protest; he said, *"Get out on the streets, or you're gonna fucking die."*

When TV interviewers asked him shitty questions, he said, *"When you say that, I hate you."*

It's what social movement academics call the emotional turn. Emotion trumps all those sensible things like "being right."

So, you could say that the reason we are going extinct is not because of all the bad guys—there are always bad guys—it's because the good guys are so crap at doing their job.

But the "climate movement" is like a family friend going to visit the Nazi household in the film *The Zone of Interest*. I'm politely suggesting it is wrong to have people die to further your problematic views.

At this point, most of you are going to turn away. It's like the bit in the Bible when the rich man comes to Jesus and says, *"Hey, so how do I get into the kingdom of heaven?"* And Jesus, speaking truth to power, says, *"You have to give away all your money."* The man walks off.

If you want to stop us from going extinct, it's very simple: you must be prepared to sacrifice. You have to disrupt to the extent the government puts you in prison for five years—just for starters. When 1,000 people do that, we will be on the foothills of not going extinct.

But just about all of you will not do that, right? Because the hell has not happened yet, right? The train has not yet actually sliced up the child's body on the track.

The biggest mistake, though, is not to decide not to act to the point of effectiveness. The most appallingly tragic mistake is to think you have chosen the easier option by walking away, like the rich man. In fact, nothing could be further from the truth.

Because what will destroy you is not the fires, the storms, the floods—like the world has never seen. Nor will it be the social effects—the economic depression, the mass migration, the social collapse, the terror of the fascists. No, what will kill you is knowing you read this article and turned away. You saw, you understood, and you did nothing. Nothing real. Nothing that actually would make the difference. Not the emotion. Not the disruption.

You disagree, right? Because you're a person of the world. You'll handle it like the man who goes to war—he can handle it too, right? And then the guy next to him gets his head blown off, and for the next 30 years, he cannot forgive himself that it was the other guy who got killed and not him.

What kills you, you see, is not the world itself but your guilt about the world. Because you are designed to care and love, and if you betray that essence of your being, you rot from the inside. And that inside, one day you realise, is all you've got.

Jesus was just trying to give the rich man a bit of good advice.

So, as Larry says, *"Get out on the streets, otherwise you're gonna fucking die."*

There are plenty of pathways to action for the few of you who are heroic enough to get this far. Look at the links.

And yes, the carbon sinks are collapsing, and the carbon budgets are fucked. And unless there is a massive turnaround, we are going extinct. But as I hope I have explained, that's not the real story

Reasons To Be Cheerful

(October 17, 2024)

State oppression will not stop climate activism, but will instead ensure climate breakdown leads to revolution.

I'm on a wing with a load of "lifers"—people serving life sentences. A while ago, one approached me and asked, "So, how long are you in for?""Five years," I said."Ah, that's not long," he replied.

I have done hundreds of public talks about our situation, and I used to remind people that we don't live in Egypt. No one is going to get their fingernails pulled out. While researching at King's College in London, I worked with a colleague who studied trade union activity in Egypt before Tahrir Square.

He would sit in a room in Cairo with a circle of guys, most of whom had scars on their faces. They had all been tortured. Shortly afterwards, the revolution happened.

The problem is not "state oppression" but those on "our side" who do the work of the "oppressor" by endlessly giving reasons why resistance "will not work." When I did strategy work for Extinction Rebellion in 2019, the biggest headfuck wasn't police oppression; it was having semi-informed "Left" journalists and other hangers-on who opined that the civil resistance model we used was no good because it was rooted in Erica Chenoweth's research in authoritarian regimes. As if people in such contexts live on a different planet! In their cynicism, they have now adopted the opposite argument: the model doesn't work precisely because we now live in an authoritarian regime.

The Egyptian Revolution in Tahrir Square, 2011

This is what I call the Aaron Bastani syndrome. Whatever happens, you refuse to leave the privilege of your journalistic gaze. When I challenged him to lead a march, he responded, "I don't think that would be wise."

This aversion to taking any real moral responsibility for the fucked-up state of the world by those who claim to speak for us is far worse than any "state oppression."

Given that I did nearly a decade of research at KCL on the relationship between levels of oppression and resistance, let me give you my one-line response to the question, "Does state oppression hinder social movements?": there is no relation. Successful movement building and popular political power are functions of internal culture—ideology, organisation, and leadership. Of course, there is a case that external repression has an impact, but if there is a signal, it is very much lost in the "noise" of other factors.

So why do our "left" and "liberal" commentators get it so wrong? Another one-line answer would be that they are materialists. They look at the world in the same way as a capitalist: more repression (cost) in, less activism (money left) out. As such, they are the most effective agents for the neoliberal regime—the obscene ideology that humanity is just stuff. Dead stuff. You shouldn't lead a march because you will lose things—clicks, reputation, liberties. The very supposed radicalism of their analysis betrays appalling reactionary politics. They are so clever; they can give you all the reasons things are so bad, all the reasons why the powerful have so much power, and why people are so oppressed—so much so that there is zero chance they can revolt. The "is" becomes an "ought." Self-fulfilling prophecy dynamics do their work very nicely. Job done. A paycheck from Wall Street is on the way. With friends like these...

On the frontline, everyone's a lot more cheerful. Because human beings, of course, are not pieces of dead stuff. They are, in fact, explosions of spirit, driven by desire, searchers for meaning, recognition, and adventure. Or, at least, when they're not being told they are too oppressed to do anything. To use the more subdued language of academia, they always have agency.

AFTERWORD

So, there you have it. those of you that have been paying attention – we are fucked. Someone just has to say it as it is. Like in a family row. Sometimes, you just have to have it out with those you love. Otherwise, what does love mean? Certainly not indifference. 'You are ruining your life, and well, that's up to you I guess.' That is not love. At least, if you are not part of the temporarily privileged Western middle class, and those that copy their ideologies.

Throughout history, humans have survived because of duty: the duty to love, to intervene. To say 'no' when the situation requires it.

It's the same with your neighbour, your tribe, your nation – humanity itself. Sometimes, you have to say 'no'. Not because it will work but because it has to be done. Because of who you are: a being embedded in this world. Without the world, you are nothing. As an act of faith, but no less important for being so. Some things, many things, if you are open to admitting it, are matters of faith.

For the past few years, I have been getting through four to five articles a day about neighbourhoods, tribes, nations – humanity itself.

What actually science is trying to say, at least to those that understand the notion of the duty to love is simply this: that what we love, our family, tribe, nation and humanity, are being and soon will be forever fucked. Let me put it this way: imagine a high fence a thousand miles long, across plains and mountains, as far as the eye can see. And on one side of that fence are millions of corpses – men, women and children – limp bodies, jagged limbs.

We know what it looks like. We in Europe cannot turn away because it has happened before.

Except this time around, it will go on forever, until there are no more millions, or billions should I say to kill, to die. Limp bodies stretching out for miles and miles and miles.

Let me cut to the chase. Maybe the most important question of our time is this: where are the people of India going to go? The Bangladeshis, the Pakistanis, the Iranians when temperatures hit 50–60°C? When the wet bulb effect kills the body – every body – in six hours? It's a science thing, right? Like being shot in the head, like being drowned in the sea, like being trapped outside at night naked at –20°C. You die. Sometimes, things are certain, and you ignore certainties at your peril.

Let me summarize my 10,000 hours of reading four to five articles a day. Over the following decades, one thousand million will be killed – that is twenty times more than those who died in the Second World War. Global gross national product will fall by 50 per cent and never recover. You're right, although these numbers come from peer-reviewed papers, they could be wrong. Which, to be honest, actually means they could be a lot worse. As a sociologist, I would say that because science exists in society and is thus subject to social influences – the conservative methodologies and political pressures – just about all the predictions are underestimates. Isn't the most common phrase in articles now, 'scientists were surprised that...'

To stop this horror of horrors, the world needs to reduce carbon emissions by around 8.3 per cent a year. During the COVID-19 pandemic, emissions fell by an estimated 4 per cent.

So, it is not going to happen. Not because it is not possible, but because it is not possible under the existing carbon regime – the present set of human-made political, social and economic arrangements. The situation can change, and it will change when the regime changes. And the name for that change is 'revolution'.

It is important to understand that revolutions are not, in the main, brought about by revolutionaries but rather by the elites in power who commit collective suicide. They decide to pursue courses of action that destroy their societies and thus themselves. They spend too much money, they engage in ruinous wars, or they destroy the natural environment – the source of all material well-being. They no longer can see reality.

The question then is not, revolution or no revolution? The question is what type of revolution: a reversion into the hell of colonialism, fascism and totalitarianism, our global history for the past few hundred years. Or something else – something new which is also very old.

You will not discover this 'something' by reading articles or books, not even this book or this afterword. Sure. You will read about solutions and plans and strategies on the page. But to be able to create a new world, to save those you love: you will only find that knowledge in the police van, in the prison cell, by passing through the long dark nights of despair. Only then you'd be presented with a decision to make: to act for God or act for yourself. There is no middle ground. It is a matter of faith. The way coming out of years of the French Resistance in World War II, the French had to decide to act for justice or act for nihilism.

I have chosen to act for justice. The key word here is 'act'. And you do not act alone – you act with knowledge. Action and knowledge are two sides of the same coin. The reason why the global climate movements have been largely unsuccessful is because they do not marry knowledge and action well.

When a scientist, politician, diplomat – all the representatives of the global carbon regime – tell you we are going over 2°C, and maybe that a billion people stand to die and do not raise their voice, do not tear off their tie in rage, don't spill tears of desperation, don't leave the podium, resign from their position, sit on the road and say 'never again'. When they do not find themselves put in a locked room... well you go back to what you were doing – like billions of others.

I have been arrested for resisting the mayhem in this world more times than I can remember. The last time was three days before writing this piece. I have been sent to prison three times in the past five years – the last time for four months for making a speech telling people what I am telling you now. I am to be sentenced for another 'offence' in a few weeks, so I may be incarcerated by the time you read this. You might need to know that I spent three years sleeping in my car and under my PhD desk doing the research that led to the founding of the Extinction Rebellion. I got suspended from my university and went on hunger strike to force it to take climate action. I got divorced so I can do this work. And I will be doing this work for the rest of my life.

What I have done and what has happened to me is nothing unusual. It is what happens when you act – really act – for justice. It is happening to millions of people right now, and hundreds of millions have struggled and died for justice in the last hundred years. This is what you need to do if you claim to love what you love and the people you love. It's time to look in the mirror.

I decided to write this text because my friend Chittranjan (Chit) said it would reach an Indian readership. I believe India is one of the countries that might lead humanity out of the abyss.

All I have written here is not new. I claim no originality. We will not learn from books but from love in action, from its agony and its ecstasy. The most important thing is 'self- purification' not the act of civil disobedience itself. Unless you understand yourself and the world, you cannot act in love, and you will just make the world worse. I am led to understand there are many in India who have turned about from such realities and are temporarily under the spell of Adam Smith – the idea that happiness is about accumulation of stuff. Of course, it is true enough, but it is now leading to hell. Because it is. Look up – look around you.

Which leads me to you and me now at this moment. I have tried to communicate that we are set within the world. There is no disembodied knowledge. You cannot sit down in a room and get the answers. But this is the conceit of modern academia rooted in Western 'rationalism' – that you write an article, you give information and argument, this is assessed (like a computer) and then people change the world.

The climate movements are rooted in this conceit. They are part of this global Westernized detached culture. It is self-serving as it avoids the necessity for commitment and sacrifice. It is, to be blunt, bollocks.

Having read what you have, you have to act. And you need to know concretely what to do. This text, this book, has to be committed and embedded, not in the comforts of intellectual gossips but in pathways to collective struggle.

Here is one. Come and join us, and you will meet people involved in some of the most effective nonviolent resistance all round the world. Then you can act. And isn't this what we are all called to do – the most we can do. Nothing more, nothing less.

I wish you all the best.

Acknowledgements

This book is the result of years of collective struggle, sacrifice, and unwavering commitment from those who refuse to stand by in the face of climate collapse. Roger Hallam cannot write this himself right now—he is in prison for his role in this fight—but his words, like his actions, are driven by a deep love for humanity and an unshakable belief in our power to resist.

On Roger's behalf, I want to thank all those who have stood with him—on the streets, in the courts, in prison cells, and beyond. To the rebels of **Extinction Rebellion, Just Stop Oil, Revolution in The 21st Century and Insulate Britain** who have put their bodies on the line, who have faced repression with courage, and who continue to fight for life itself—this book belongs to you.

To those who have supported Roger during his imprisonment—through letters, solidarity actions, and simply refusing to let him be forgotten—your care and defiance mean everything. To his family and close friends, thank you for holding him through the storm.

And to the great lineage of movement builders, thinkers, and revolutionaries whose insights shape these pages—your work fuels the next generation of resistance.

This book is a call to action. It is written in the spirit of all those who dare to dream of revolution and are willing to fight for it. Roger may be behind bars, but the struggle continues.

Robin Boardman, on behalf of Roger Hallam 2025

Printed in Great Britain
by Amazon